MOULD DESIGN USING HYDRAULIC PRESS

For Refractory Bricks – The practical way to designing moulds

SHEOJEE PRASAD

Rev 1.0 dated 15-Jan-2017

ABOUT THE AUTHOR

Sheojee Prasad has 40 years of experience in Pattern design, Pattern making, Mould design and moulds manufacturing for production of refractory shapes. With experience working in Tata Refractories Limited (now known as TRL Korasaki Refractories Ltd), he brings in his expertise through managing pattern shops and designing wooden and steel moulds in the form of this book.

This book is a practical guide for refractory shape making using hydraulic press. It leverages Sheojee's experience in Pattern Shop, Wooden Mould shop, Steel Mould shop, Machine shop, Flow Control Mould shop, Mould design and Refractories lining design.

ALSO BY SHEOJEE PRASAD

1. **Learn Critical Aspects of Pattern & Mould Making in Foundry Vol. 1 (Volume 1)**

http://amzn.com/1519449429

2. **Pattern & Mould Making in Foundry – Vol.2 (Practical Exercises)**

http://amzn.com/1530573408

ACKNOWLEDGEMENTS

In dedication to my sister who had a great influence throughout my life.

CONTENTS

Why I wrote this book

I: Introduction to hydraulic press, refractories and mould

Working principle of hydraulic press
Refractories and its application
Refractory materials
Classification of refractory
General product of refractory industries
Product range according to raw material and composition
Sketch of few brick shapes -3D

II: Action plan in the formation of shapes

Manufacturing of refractories shapes
Elements of process design and planning
Mould
Bulk density
Process design
Contraction and expansion allowances

III: Mould design

Mould design requirements
Category of moulds on the basis of intricacy and materials
Category of moulds on the basis of pressing mechanism
Accessories of Hydraulic press moulds and 3D sketch

IV: Hydraulic press Mould design requirements

Collection of information to design mould
Specific Pressure
Press Configuration and working parameter

V: Mother mould design

Mother mould for hydraulic press
Organs of mother mould with 3D sketch
Design of mother mould
Outer shape of mother mould

Pocket design in mother mould
Edge pressing and flat pressing explanation with 3D sketch
Cavity design in mother mould
Wall thickness calculation
Filling depth and height of mould

VI: Cavity size in Mother Mould and tap holes

Absolute cavity size calculation
Absolute cavity size when length and width of mother mould, pressing capacity and a specific pressure is known.
Decision on length and width of cavity in mother mould
Calculation for No of cavities and exercise
Decision on direction of mould cavity with respect to brick size
Calculation of mother moulds walls thickness with respect to brick length and specific pressure.
Exercise
Direction of mould cavity and formula for calculating cavity size in mother mould
Design of tap holes at top and bottom surface of mother mould

VII: Lock plates and its application

Design of lock plates
Organs of lock plates
Different systems of lock plates
Representation with 3D drawings

VIII: Liners, loose piece and packing design

Design of side and end liners
Design considerations
Loose pieces
Packing plates
Packing size calculation

IX: Plunger design

Top and Bottom plunger
Organs of plungers, description with 3D drawings
Plungers design to suit operating parameters of press

X: Press configuration and related calculations

XI: Design of dies and plunger holding plates

Working parameter of press
Calculation of maximum brick thickness that can be pressed
Corrections in configuration of press parameter

XII: Brick shapes for mould design

XIII: Practical guides on grouping

XIV: Edge pressing mould design for making key bricks

XV: Mould design for side Arch and End arch bricks.

XVI: Mould design for coke oven brick

XVII: Mould design for dome brick

WHY I WROTE THIS BOOK?

The opportunity to help and guide a new generation of people working in the mould design and refractory shape industries was something that I wanted to explore and contribute to. Having worked and learned through colleagues during the process, I wanted to pass on my valuable insights to you.

I have also authored two books, one providing a basic foundation on pattern and mould making with the 2nd book providing practical exercises to guide practitioners.

This book takes you to through important expects of mould design and will inspire those who are willing to work in refractories shape making organisations; it is also a guide to supervisors of production and maintenance.

Chapter - 1
Introduction

Working principle of hydraulic press:

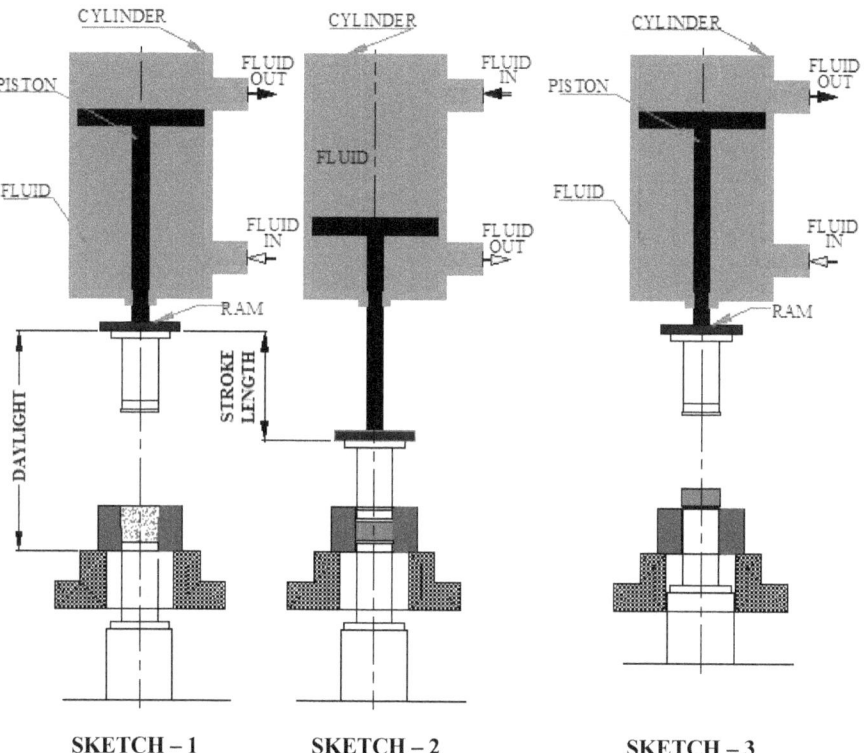

| SKETCH – 1 | SKETCH – 2 | SKETCH – 3 |

A hydraulic press is working on the principle of Pascal's law and it relies on differences in fluid pressure. The basic working principle has been explained with a sketch in three stages.

Sketch – 1

When fluid is pumped into the cylinder below the piston and simultaneously fluid is pumped out from the top it causes an increase in fluid pressure under the piston and decrease above the piston. The top ram along with plunger and die will move up creating gap above mould surface for charging the brick mixture. The cavity filled with the mixture is visible in the sketch.

Sketch – 2

In the next cycle of operation, the fluid is pumped out from below the piston and simultaneously fluid is pumped into the cylinder from the top. The higher pressure of fluid above the piston and simultaneously decrease under the piston will move ram with the plunger and die downwards with greater force to press the brick mixture in the mould cavity. The compressed brick mixture to required size in between top and bottom dies is indicated in the sketch.

Sketch – 3

In the third cycle of operation, fluid is pumped into the cylinder below the piston and simultaneously fluid is pumped out from the top, it causes an increase in fluid pressure under the piston and decrease above the piston. The top ram along with plunger and die will move up leaving pressed brick on the bottom die surface in the mould cavity. Finally, ejection stroke is applied from ejection cylinder working on the same principle. The ejected brick on the bottom die has been drawn above the mould top surface.

The detailed knowledge of press component and the working parameter will be available in maintenance manual supplied by press manufacturer.

Common specification of hydraulic press:

Model No, Force (Tonnes), Work area – front to back, Overall width – left to right, overall height, Clearance between columns – front to back, Daylight - maximum opening, shut height- daylight minus maximum stroke length, Bed to floor height.

The main structure of the press consists of the column, cylinder and working surface. Cylinder assembly consists of a cylinder, piston, ram, packing and seats.

Stroke control: The stroke length can be set for any distance within the stroke limits of the cylinder by the control panel and switches. The adjustment includes top stroke, pre-slowdown point, daring and bottom stroke.

Daylight: It is the vertical clearance from the top of the bolster to the underside of the ram in its maximum up position. It is also mentioned as the vertical distance from the top of the ram to the floor.

Bolster: It is a plate or structure mounted on the bed of the hydraulic press. It may have a removable system.

Bed: It is a flat stationary machined surface that supports mother mould. It has a cavity in the centre for free movement of plungers and associated parts.

Refractories:

Refractories are materials that have very high fusion temperature. It is used in various sizes and shapes to save the structure of metals from deformation and melting down during heating process for specific uses.

All sorts of heating equipment's shell or structures are lined with refractory materials to maintain stability, preserve heat and prolong the life while under exposure of intense high temperature during operation.

Common area of application

- **Steel plants and foundry:**
    1. Blast furnace
    2. Converter
    3. Ladle
    4. Tundish
    5. Electric arc furnace
    6. Cupola Furnace
- Glass Tank (Glass Industries)
- Limekiln
- shaft Kiln
- Chemical plants
- Cement plants
- Power plants
- All other heating equipment that is subjected to high temperature during operations.

Refractory Materials

- Fireclay
- Oxides of Alumina
- Silicon(Silica)
- Magnesium
- Dolomite
- Zirconia
- Silicon carbide
- Graphite

Classification of refractories on the basis of stability against slag and prevailing atmosphere in furnace

- **Acidic refractories:** The shapes are not affected by acidic materials
- **Neutral refractories:** These are chemically stable and used in an area where slags and atmosphere are either acidic or basic.
- **Basic refractories:** These are used where slags and atmosphere are basic. They are stable to Alkaline but could react with acids.**Fusion temperature of refractories**

Grade of refractories	Fusion temperature
Normal refractories	1580 -1780^0 C
High refractories	1780 -2000^0 C
Super refractories	More than -1780^0 C

General products of Refractory Industries

- **Shape bricks**
    1. Standard size bricks
    2. Nonstandard-size bricks
    3. Shape bricks such as for coke oven shapes
    4. Special shape bricks
- **Monolithic**
- **Ramming mass**
- **Gunning mass**
- **Mortar**

Product range as per raw material and composition

- Silica
- Basic
- Fireclay
- High alumina
- Dolomite
- Special refractories.

Few Brick shapes:

Key Bricks: There are two types of key bricks,

- Key along length having an equal taper on both sides starting from one end to another end.

- Key along width having equal taper on both ends starting from one side to another side

Side Arch brick: It has an equal taper on both faces starting from one side to another side. Both faces must be level and warpage-free. (See given figure). In this case, the taper is along the side.

End Arch Brick: It has an equal taper on both faces starting from one end to another end. Both faces must be level and warpage-free. (See given figure). In this case, the taper is along the length.

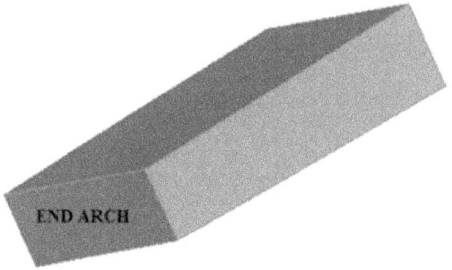

Radial bricks: These bricks are bounded by two radial lines and two arcs. When the radius of Arc is very large, the radial line is removed and two ends are joined with the chord. (See the figure drawn here)

Dome bricks: It has four sides taper starting from one end to another end. See figure for better knowledge.

Coke oven Brick: This shape has been drafted to explain the process of mould design. It is not for application. There are various shapes and sizes of coke oven bricks made by refractory lining designer.

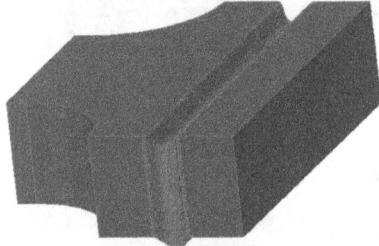

Skewback brick: It is used as support at the end of the wall for the construction of arches.

Chapter – 2

Action plan in the formation of shapes

Manufacturing of refractory shapes

The manufacturing of brick shape needs special effort such as

- Process design
- Planning.
- Selection of raw material

Elements of Process design and planning

- Mould making
- Setting the composition
- Mode of forming shapes to refractory materials, to suit the application need at customer's end, such as machine moulding, hand moulding, precast shape etc.
- The bulk density of shape to suit the application.
- Selection of press
- Dry chamber processing.
- Processes of firing
- Inspection

Mould:

It is an object having inner and outer shapes. The inner shape is formed by mould maker according to brick shape and size with necessary allowances. In the case of simple shape, it will be fully or partially negative shape of brick to be made. The outer shape is called mother mould. The size and shape of the outer body are designed to suit the press requirements. The outer and inner shapes are also governed by various other factors that will be clear in due course.

Bulk density:

The bulk density is defined as the dry weight of soil per unit volume. Bulk density considers both the solid and the pore space whereas particle density considers only the mineral solids. The bulk density is important because it tells about the porosity of solid bricks. It is expressed in percentage. The structure of refractory having higher bulk density will be denser resulting in better resistance to chemical attack, decreased metal penetration and better abrasion resistance.

Process design:

It is the responsibility of Research and Development Laboratory to set the composition to suit the customer's requirement. The person concern will take decision on

- Suitable composition
- Expansion or contraction allowances
- Setting, drying and firing
- Mode of forming the shape

Expansion and contraction allowances:

The expansion and contraction of bricks on firing depends on the quality of the composition of the brick mixture. The brick that expands on firing expansion allowance is added on linear dimension. It is negative allowance which is subtracted from basic measurement given in drawing. It is established by research and development laboratory.

The brick that contracts on firing, contraction allowance is added on linear dimension. It is a positive allowance that is added on nominal basic drawing size. It is also determined by research and development laboratory. These allowances are expressed as a percentage.

Planning:

The person's concern will consider

- Delivery period
- Manufacturing processes that is
  1. Selection of mode of pressing: Hand moulding or Press, Precast or Ramming
  2. Mould design and mould making time
  3. Brick or shape manufacturing time
  4. Setting, drying and firing time
  5. Inspection

The planning is very important and knowledge oriented work. The planner must have knowledge of manufacturing process and <u>clear concept of engineering drawing.</u>

Chapter- 3

Mould design

The mould design is very important in the formation of refractories shape. The critical shapes also can be made on press with a suitable design of the mould. The mould designer must have a clear concept of engineering drawing. The minor mistakes in mould design may cost heavy on product and finally on the company.

Mould design requirements:

- Bulk Density of shape
- No of pieces to be moulded
- Drawing of shape to consider ease of ejection from the press.
- The quality of brick with expansion allowance or shrinkage allowance to be considered on mould dimensions.
- Assembly or non-assembly shapes
- Mode of inspection by customer
- Pressing mechanism.
- Press capacity and Specific pressure to be applied, it is essential for taking a decision on no of cavities.

Category of moulds on the basis of mould material, intricacy of shape and no of pieces required:

- **Steel mould:** These are made for the higher production of shapes that can be easily ejected from the mould cavity.
- **TC Liner Mould:** The inner surface of the mould cavity is built with tungsten carbide. It is used for mass production.
- **Wooden mould:** Wood is used to make the complicated shape of bricks.

Steel moulds:

The various grades of steel with hardness are used for making steel moulds. The factors involved for selection of steel, decision on heat treatment and hardness depends on

- No of bricks required
- B.D. of bricks to suit the application
- The quality of bricks and abrasiveness of composition.

The shapes that can be easily ejected from mould are planned for the press.

Category of moulds on the basis of pressing mechanism;

- Hydraulic press mould
- Mechanical press mould

Accessories of Hydraulic Press mould:

- Mother mould
- Sideliner
- End liner
- Loose pieces
- Packing plates
- Packing seems
- Top Plunger
- Bottom Plunger
- Plunger holding plates (Top and Bottom)
- Top Die
- Bottom Die
- Lock plates (Top and Bottom)
- Allen bolts

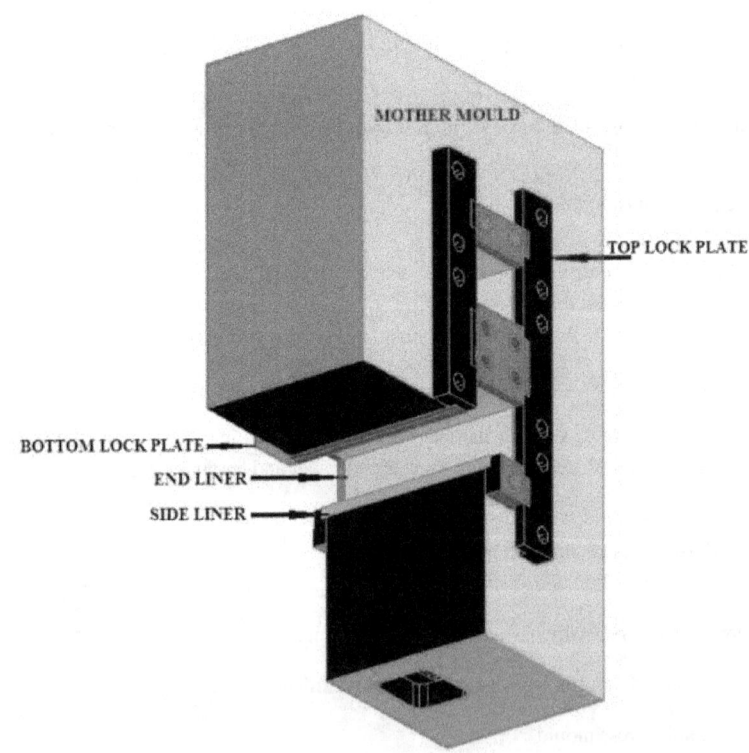

DOUBLE CAVITY MOULD ASSEMBLY

Packing plates are on the back side of liners. There is no loose piece in this mould assembly. The loose piece will be indicated in manufacturing assembly drawing.

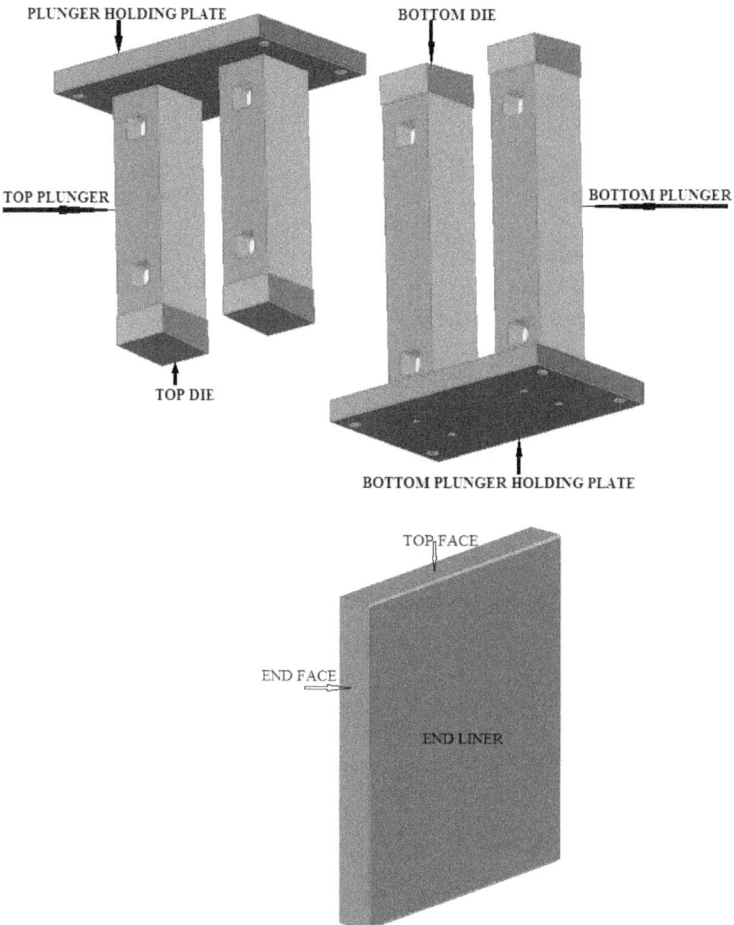

The detailed design of side-liners and end liners shapes under required condition and brick shapes have been designed and explained in preceding chapters.

Chapter – 4

Hydraulic press mould design requirements

Mould design requirement for Hydraulic Press:

Collection of information:

- Specific pressure required for a particular composition
- Available press in the plant
- Operating parameters of press
    1. Pressing capacity
    2. Operating system of press: hydraulic or mechanical
    3. Pressing stroke length
    4. Ejection stroke length
    5. Gap between mould fixing table and pressing cylinder
    6. Charger capacity and stroke length
    7. Height of charger
    8. Automatic charging or manual
    9. Mould clamping system details on mould fixing table
    10. Brick gripping capacity of the gripper.
    11. The stroke length of floating mould table.
- The detail drawing of shape to be made.
- Facilities available in the workshop.
- Grinding allowance needed on respective surfaces of mould accessories.

Specific pressure:

It is the pressure exerted by die face on the surface of the brick mixture inside the mould cavity to have required bulk density. It is expressed as a tonne per square centimetre. It is used for calculating the number of the cavity that can be made in mother mould considering pressing capacity of the press.

Specific pressure in Metric Unit (S_P) = Press Tonnage (pressing capacity) / pressing surface area of mould cavity top or (Die face)

S_P = Capacity in tons * 1000 / Area of die face (mm^2), it will be in Kg / mm^2

S_P = Capacity in tons * 1000 / Area of die face in cm^2, it will be in Kg / cm^2

Configuration and working parameter of a particular press: It has been given to explain the mould design process. The basic principle of mould design will remain same but plunger size, pressing system, ejection and removal process of brick from the press will vary according to operating parameters. The no of cavities will also change.

Daylight = 3390 mm. Stroke length of top ram =800 mm

The stroke length of mould table =1250 mm. Charger height = 380 mm. Distance of

Distance of plunger holding base from floor = 368, Distance of mould table from floor = 483

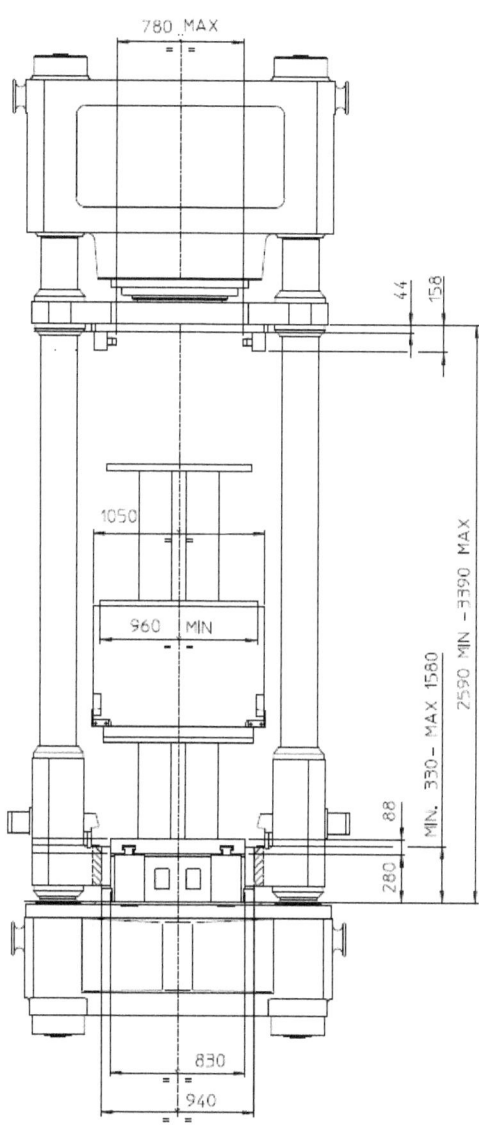

PRESS CONFIGURATION DETAIL

Chapter-5

Mother mould design for hydraulic press

Mother mould for Hydraulic Press:

The mother mould is the main body that holds liners, loose pieces and other accessories to form the mould shape which is partly or fully negative shape of a brick. The mother mould maintains the rigidity and shape of the mould during pressing or formation of bricks. Its shape consists of length, width, height and a cavity in the centre having length and width to hold liners, packing etc.

Organs of mother mould:

- Mainframe having length, width and height (thickness)
- Cavity in the centre to hold liners and packing
- Pockets in two sides to fix on mould fixing table
- Tap holes on top and bottom face to fix lock plates with Allen bolts for holding liners.
- Tap holes in the pocket base to fit loose pieces for ease of holding by a hydraulic clamp on mould fitting bed (Table).
- Plane hole in the pockets for inserting necessary bolts to fix mother mould on the mechanical press.
- Drill holes on the both end face for inserting pins to lift with an overhead crane.
- Circular bore at four corners of the cavity in mother mould: It is an essential feature of design in mother mould. It must have smooth surface throughout depth without any scar mark on the circular surface. If it is maintained there will not be any point of stress concentration due to the horizontal component of compressive when the force applied on mixture inside the mould cavity.

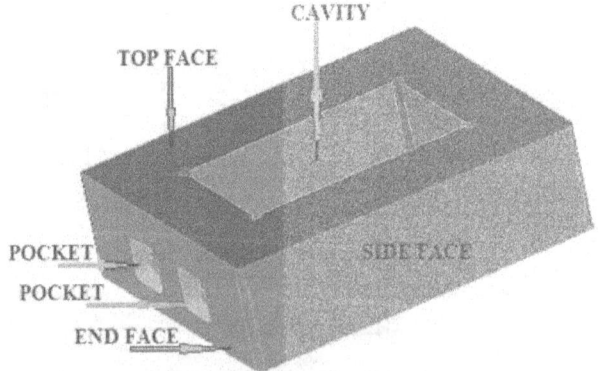

MOTHER MOULD WITH POCKETS AND CAVITY

The top face and bottom face of mother mould will have tap holes for the fitting top and bottom lock plates. The right side end face also has two pockets same as in left end face. These features will be designed in exercises.

Design of Mother Mould

The Mother mould is designed on the basis of following information and considerations:

- Holding facilities available on press's mould fitting table, Hydraulic clamping or Mechanical clamping.
- List of items for the formation of the group.
- Pressing thickness (brick thickness)
- Edge pressing or flat pressing.
- Filling depth
- The design of pocket and holding mechanism.
- No of cavities for moulding   brick sizes or shapes
- Pressing capacities of press (press tonnage)
- Maximum length and width that can be accommodated on mould fixing table of press.- It will decide outer size and shape of mother mould
- Quality of bricks to be produced
- The specific pressure required to form shapes with required B.D.
- The strength of the mother moulds to resist pressing load.

Outer shape of mother mould:

It is decided on the basis of

- The available area on mould fixing table.
- Clamping tap hole distance and tap size in case of mechanical clamping system
- Maximum and the minimum distance between hydraulic locking clamps. Maximum distance in between the left and right side hydraulic clamps when in a neutral position. The minimum distance is in between when the hydraulic clamps move forward under hydraulic pressure to hold the mother mould. we can also say, It is the distance between two arms of grippers and the stroke length that will hold the mother mould.
- Transmission of pressure on the wall of the mother moulds when pressure is applied to cavities. Required cross section of side and end walls.
- Pressing capacity of press.
- The available area in between columns.

The study of press manual supplied by press manufacturer will give you detail knowledge of gripper design its use and operating parameter.

Pocket design in mother mould:

There are two categories of pockets in mother moulds

1. Mechanical clamping with bolts
2. Hydraulic clamping with hydraulic cylinder and clamps.

Basis of Pockets design in mechanical clamping system

- Size of tap hole on moulds fixing bed or table
- The size of the bolt.
- type of bolt (Allen or Hexagonal)
- The width of pocket is decided on the basis of size and type of bolts to be fitted considering movement and gap required for tightening the bolts with Allen key or spanner.
- The height of pocket is decided on the basis of length of bolts that are to be fitted.

Pocket for Hydraulic clamping system:

To design zero defect pocket consider

- Linear movement of grippers
- The distance between two ends of gripper after and before expansion.
- Size and shape of the gripper.
- Central distance between left and right or front and back grippers

The above all information is provided by press supplier. The drawing with the size of gripper holding plate fitted in pockets is also provided in the manual.

Design of cavity in mother mould:

The cavity size in mother mould is very important for assembly of the liners to form mould cavities for the production of the bricks. It is designed after following activities:

- Make a grouping of items that you want to produce on a particular press.
- Access the manufacturing period of items on order considering possible no of cavities formation.
- Calculate packing thickness in between two liners to form the middle walls in case of multi-cavity mould design.
- Consider the width of the middle arm of the gripper and its movement in left and right sides for deciding middle wall thickness.
- Maximum opening between grippers arm
- Decide side and end packing pieces in the back of liners.

Calculation of cavity size in mother mould:

The cavity size in mother mould is calculated on the basis of

- Specific pressure to calculate absolute cavity sized and no of mould cavities.
- The outer periphery of brick.
- The maximum length of brick in groups.
- Maximum width of brick in groups
- The width of gripper arm.
- Pressing capacity (Tonnage of press)

- Calculated no of mould cavities for the production of bricks.
- A group of items that are on regular production or expected to be.
- Liner and packing thickness

Importance of Grouping:

- It will provide an opportunity to have maximum utilisation of mother mould.
- It will help in deciding common **cavity size**.
- It will help in deciding the height of mother mould considering **filling depth.**
- It will help in making a separate group for **edge pressing and flat pressing**.

Filling depth:

The loose refractories mixture will need more volumetric space than pressed brick shapes. An extra depth of 80% to 100% is considered on brick thickness to accommodate the loose mixture in the cavity. This extra consideration is called filling depth. The total filling depth will be brick thickness + 80 % of the brick thickness.

Suppose your required brick thickness after pressing is 100 mm.

Filling depth will be 80% of 100 = 80 mm, Total filling depth will be = 100 +80 =180 mm.

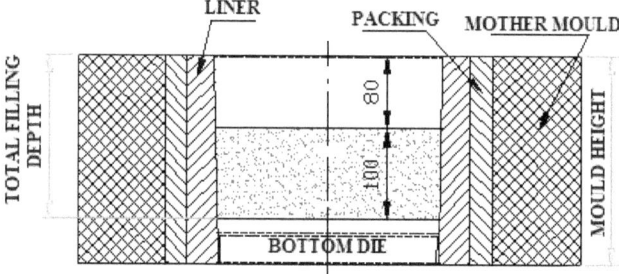

Height of mother mould:

The height of mother mould is composed of total fill depth + thickness of bottom die. The bottom die must be inside mould cavity during mixture charging operation. It is not necessary to have full thickness of bottom die inside mould cavity but 50% must be inside.

The height of mother mould in above case will be = 180 + 50 = 230 mm.

Suitable Height of mother mould on press:

Factors involved in taking decision on height are

- Maximum and a minimum thickness of bricks that are to be pressed in a group.
- Total fill depth
- The gap between mould fixing table and top ram of the press.

- Height of charger
- Height of gripper
- Thickness of bottom dies plate that is to be accommodated inside mould cavity
- Thickness of top and bottom lock plates
- Stroke length of floating mould table
- Stroke length of pressing cylinder.

Edge pressing:

All rectangular body have 6 surfaces, 2 faces, 2 end faces, 2 side faces. When mould cavity is bounded by length and thickness of brick and pressure is applied on mixture filled in it, is called edge pressing. Generally, length and thickness are considered for edge pressing but width and thickness also can be considered for edge pressing when the first option is not possible. The bricks must be easily ejected after pressing operation from the press in any option.

EDGE PRESSING SURFACE

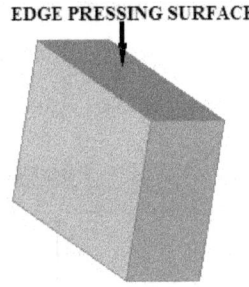

Flat pressing:

When mould cavity is bounded by length and width and pressure is applied on mixture filled in it, called flat pressing.

FLAT PRESSING SURFACE

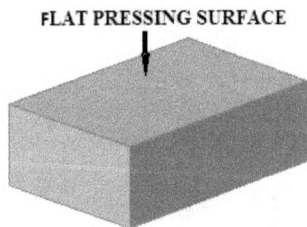

Advantage of edge pressing over flat pressing:

- More production
- Better quality having improved B.D. and uniformity

Category of group:

- The separate group for individual quality.
- Common mother moulds for assembly of different sizes of the bricks.
- The variable pressing thickness in the same mould cavity.

Assessment of manufacturing period:

- Customer satisfaction due to timely delivery of product
- Availability of mother moulds for next assembly

Benefit of calculating middle wall in between two cavities (two liners and middle packing):

- Prior assessment will have the benefit of smooth movement of gripper middle arm.
- Will provide required grip on bricks while taking off from bottom dies.

Benefit of considering width of middle arm of gripper and its movement:

- Will provide appropriate clearance between the arm and two bricks sides.
- No bricks damage during forward movement to grip the bricks.

Benefit of side and end packing plates:

- Will accommodate bricks of other sizes.
- Maximise utilisation of cavity

Chapter – 6

Mother mould cavity and tap holes

Mother mould cavity and mould cavity:

Let us first understand the difference between mould cavity and mother mould cavity. The mother mould cavity comes first then mould cavity.

Mother mould Cavity: It is opening in the centre of mother mould having length and width in the centre of mother mould. The depth or height is equal to the height of mother mould. It holds liners and packing plates.

Mould Cavity: The mould cavity is formed in the mother mould cavity by the assembly of liners and packings. It is used for making refractory bricks. The formation of mould cavity may be a single cavity or multi-cavity depending upon press capacity, specific pressure and facilities to take off bricks from the press.

Absolute Mould Cavity size in mother mould with specific pressures:

The absolute mould cavity here refers to the net surface area of die face or open face of mould bounded by the liners.

The specific pressure required for pressing the brick depends on the bulk density of a brick. The absolute cavity size is calculated on the basis of required specific pressure to have bulk density.

Sometimes the specific pressure is mentioned as kg / square cm. in this case, divide specific by 1000 to convert it to Tonne / square cm.

When it is given in Tonne / square cm multiply it by 1000 to convert into kg / square cm

The unit of press capacity and specific pressure should be same to get absolute cavity size.

Calculate the absolute cavity size in mother mould for 0.8, 1, 1.6 and 2 tonnes / square centimetre specific pressure for a press of 1000 tonnes pressing capacity.

Absolute mould cavity with 0.8 specific pressure / square cm

Absolute cavity surface area with 100 % utilisation and 0.8 tonne per square centimetre specific pressure = 1000/0.8 =1250 square centimetre.

Absolute mould cavity with 1.0 specific pressure / square cm

Absolute cavity surface area with 100% utilisation and 1.0 tonne per square centimetre specific pressure = 1000/1.0 =1000 square centimetre.

Absolute mould cavity with 1.6 specific pressure / square cm

Absolute cavity surface area with 100 % utilisation and 1.6 tonne per square centimetre specific pressure = 1000/1.6 = 625 square centimetre.

Absolute mould cavity with 2.0 specific pressure / square cm

Absolute cavity surface area with 100 % utilisation and 2.0 tonne per square centimetre specific pressure on die face = 1000/2.0 = 500 square centimetre.

The absolute cavity size means a top surface area of cavity bounded by length and width or die face that will be subjected to pressing load. The cavity in mother mould has to be increased by the addition of total thickness of liners and packing plates. The no of the liners and packing plates are to be calculated according to no of cavities.

The above calculations indicate that the increase in specific pressure would reduce surface area coming under die pressure.

Calculation of the Absolute cavity size in mother mould under given condition such as:

- Length and width of mother mould
- Pressing capacity
- Specific pressure

Calculate absolute length and width of the cavity for a press having 1000 tonnes pressing capacity, the length of mother mould 1050 mm and width 910 mm.

Specific pressure = 1 tonne / square cm = 1000 kg / square cm

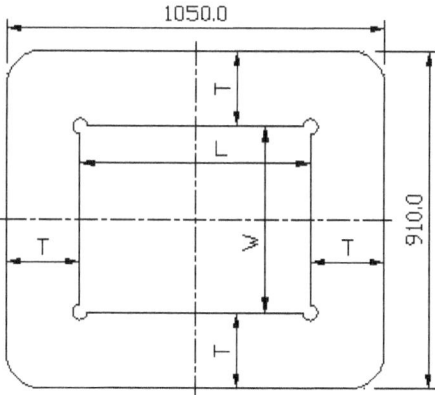

- T = 0.01* L * √ P$_S$ in cm
- T = wall thickness in cm
- L = Absolute cavity length in centre (cm) including liners and packing
- W = Absolute cavity width in centre (cm) including liners and packing

- Absolute cavity length + 2* T = Mould length (considering the above figure).

  L = Mould length – 2*T = 105.0 – 2*(0.01* L * $\sqrt{P_S}$ in cm)

  L= 105.0 – 2* (0.01* L * $\sqrt{1000}$) = 105.0 – 0.02*31.62 L

  Or L + 0.6324L = 105.0

  Or L = 105.0 / 1.6324 = 64.32 cm =643.2 mm

  T = (1050 – 643.2) / 2 =203.4 mm (say 205 mm).

  As equal pressure will be on four walls of mother mould, wall thickness should be equal.

  W = 910 – 410 =500

  Maximum cavity size in mother mould can be 640 long and 500 wide.

Calculate absolute cavity length and width in mother mould having length = 1300 mm and width = 1300. The pressing capacity is 1600 Tons and required specific pressure is 1.6 tonne / squire cm

- $T = 0.01* L * \sqrt{P_S}$ in cm
- T = wall thickness in cm
- L = Absolute cavity length in centre (cm)
- W = Absolute cavity width in centre (cm)
- Absolute cavity length + 2* T = Mould length

  L = Mould length – 2*T = 130.0 – 2*(0.01* L * $\sqrt{P_S}$ in cm)

  L= 130.0 – 2* (0.01* L * $\sqrt{1600}$) = 130.0 – 0.02*40 L

  Or L + 0.80L = 130.0

  Or L = 130.0 / 1.8 = 72.22 cm =722.2 mm Take it as 720

  T = (1300 – 720) / 2 = 290 mm

  As equal pressure will be on four wall of mother mould, wall thickness should be equal.

  W = 1300 – 580 = 720 mm

  Maximum cavity size in mother mould can be 720 x 720 mm

Decision on formation of mould cavity for production of bricks in mother mould:

- The study of gripper's detail working parameter is very important if gripper is to be used for taking out brick after pressing and ejection.
- Holding capacity of gripper: How the gripper will hold the brick for taking out from the press after ejection operation. Can it hold bricks lengthwise or width wise? It will depend on maximum and the minimum distance between two gripper's arms. Forward and backwards movement of the gripper are important. The length of the arm is also important.
- No of cavities decided according to pressing capacity of the press and required specific pressure. It is an essential factor in calculating length and width of the cavity in the centre of mother mould.
- The wall thickness of mother mould on both ends of the length and width + cavity size with a thickness of liners and total packing thickness must accommodate on

mould table. It is the equation of length and width of mother mould which is decided and designed by press manufacturer.

- The formation of mould cavity size in mother mould is also decided on the basis of
  1. A group of items to be moulded considering length, width and thickness of bricks.
  2. Maximum and minimum specific pressure required for different qualities of bricks. It will provide the maximum and minimum absolute surface area of die face and cavity size.
  3. Absolute surface area which is calculated on the basis of specific pressure.
  4. Length and width of mother mould that is specified by press manufacturer for a particular pressing capacity.
  5. Length and width of brick: Formation of mould cavities for making bricks is decided according to brick length and no of cavities. When the assembly of liners for longer brick size width wise will reduce the cross section of mother mould making it liable to crack during pressing, it should be laid lengthwise even in the case of multi-cavity design.
  6. Compare theoretical and physical cross section. The physical cross section should be higher than theoretical.

Calculation for number of mould cavities:

It is calculated on the basis of pressing capacity of a hydraulic press and required specific pressure to have a bulk density of a brick. First, calculate absolute cavity size and surface area of die face of one cavity. Divide absolute cavity size by the surface area of one die face.

Absolute Cavity Size (cm^2) = Press tonnage / required specific pressure in tonne

If press tonnage is 1000 tonnes and a specific pressure is 0.8 tonnes / Sq. cm, then absolute cavity Size or absolute die surface area will be =1000/.8 = 1250 Sq.cm

No of cavity = Absolute cavity top surface area (Sq.cm) / Surface area of die face or open top surface area of one cavity in sq.cm.

Exercise:

Find the no of the cavity that can be made in a mother mould for brick size 230 x 115 x 76, pressing thickness is 76 mm and press capacity is 1000 tonnes. Consider specific pressure of 1 tonne / square centimetre to have required bulk density.

In this case, the die surface area or open surface area of cavity top surface will be 230 x 115 = 23 x 11.5 square centimetre = 264.5 square centimetres.
Pressure per cavity with 1 tonne / square centimetre specific pressure = 264.5 x 1 = 264.5 tons
No of cavities = 1000/264.5 = 3.7 No of cavities with 97% utilization=3.7*0.97 = 3.59. We can design 3 cavities mould for this press to have required size and B.D.
This rule is only applicable for producing solid bricks

Calculation for no of cavities for well blocks and checker like bricks:

It will be an absolute surface area of die face. It is calculated by subtracting area of holes made on die face for inserting pins during pressing stroke from the overall surface area of the die or top surface of the mould cavity. The detailed design will be available in the next volume.

Taking off bricks from Die face:

Brick can be taken off with the help of grippers: available facilities with press must be considered.

- Grippers have provision to hold, lift and drop in the front area of mould cover plate. In this provision, the bottom dies surface should be slightly above mould surface.
- The gripper may have provision to push the bricks in front of mould cover plate. In this case, the bottom dies surface must be aligned with the surface of mould cover plate's surface.
- Taking off bricks from press also depends on a pressing system of bricks. The key bricks compressed in between dies in edge pressing can have to be lifted up and placed on platform matching with key taper, otherwise, has to be handled manually.

Mother mould's wall thickness with respect to brick length and specific pressure:

- $T = 0.01 * L * \sqrt{P_S}$ cm
- T = wall thickness in cm
- L = maximum length of mould box opening (cavity in centre) in cm
- X = length of brick in cm
- Y = width of brick in cm
- N = No of brick per stroke (No of cavities formed by liners)
- P = Total pressure in Tonnes
- A = Total surface area of bricks in cm^2 = $(X * Y) * N$ cm^2
- P_S = Specific pressure kg / cm^2
- $L = (X \text{ or } Y * \text{No of Cavities}) + \{(\text{middle packing thickness} * (\text{No of cavities} - 1)\} + (\text{liner thickness at top} * \text{No of cavities} * 2) + (\text{End packing thickness} * 2)$
- $P_S = P / A$ (kg / cm^2)

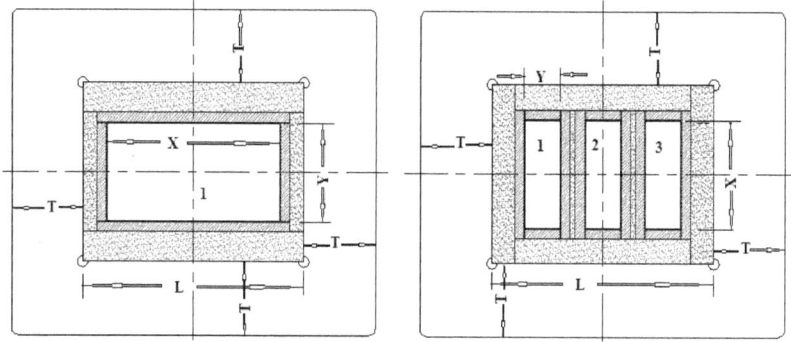

Consider left side figure and find 'T' for 1600 tonnes press

If X=50 cm, Y=25 cm

Total surface area of brick's pressing surface = 50* 25 = 1250 cm^2

P_S (Specific pressure on brick surface) = (1600*1000/ 1250) kg / cm^2 = 1280 kg / cm^2

L = 50 + (3*2) =56, where 3 cm is the liner thickness.

Packing is also an essential element of the mould assembly. Let us consider 20 mm thickness of packing on both sides of length and width.

In this case L = 56+2*2 =60

T =0 .01* 60* $\sqrt{1280}$ =0.01*60*35.8cm = 21.48cm. =214.8mm

Direction of mould cavity and formula for calculating cavity size in mother mould:

The formation of the mould cavity, its direction and number of cavity depends on facilities available to take of bricks from the press. The mould cavities may be arranged in a column, rows or in a combination of rows and columns if the press is equipped accordingly with facilities to take off bricks after ejection from the press. The given diagrams represent the formation of the mould cavity in columns (width wise) and in rows (length wise).

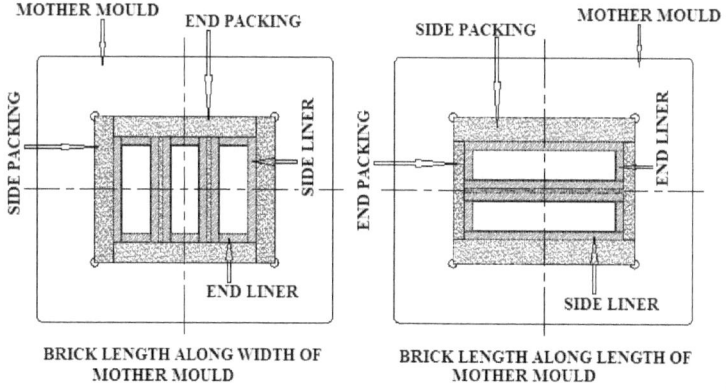

BRICK LENGTH ALONG WIDTH OF   BRICK LENGTH ALONG LENGTH OF
        MOTHER MOULD   MOTHER MOULD

When length of brick is along the width of mother mould:

Cavity width = Brick length + (liner thickness at top*2) + (side packing thickness*2)

Cavity length = (Brick width*No of Cavities) + {(middle packing thickness* (No of cavities − 1)} + (liner thickness at top* No of cavities*2) + (End packing thickness*2)

When length of brick is along the length of the mother mould:

Cavity length = Brick length + (liner thickness at top*2) + (end packing thickness*2)

Cavity width = (Brick width*No of Cavities) + {(middle packing thickness* (No of cavities − 1)} + (liner thickness at top* No of cavities*2) + (side packing thickness*2)

Note: Packing thickness is variable. It is to adjust mould cavities within tolerance. Liner thickness is also variable.

Note: Provide pockets of the same size and at the same location on side face as on end face of mother mould to facilitate uniform mixture feeding and taking off bricks after ejection with the gripper. It will facilitate rotation of mother mould if brick cavity length is laid lengthwise. Before implementing you must see and measure.

- The cavity size in mould fixing table
- The possibility of rotating mother moulds and fixing with the hydraulic clamp

Exercise:

Calculate cavity size and direction of mould cavity in mother mould for making bricks of size (1) 500 * 373 * 100 (2) 300 * 150 * 100 (3) 550 * 150 * 100

Suppose outer size of mother mould that can be accommodated on press have, Length = 1050, width = 910 and Specific pressure = 0.8 tonne / cm^2

The pressing capacity of press = 1000 Tonnes.

Maximum cavity size in centre of mother mould with 1.0 tonne / cm^2 specific pressure = 1000 / 1.0 = 1000 cm^2

The absolute cavity size in the centre of mother mould will be 1000 cm^2.

The length and width of the cavity will depend upon the size of bricks to be moulded. Generally, maximum length and width are considered in the group to accommodate other sizes in between.

Suppose maximum size of brick to be made is 500 X 375 X100

Pressing surface of brick if flat pressing is considered = 500X375 = 1875 cm^2. It is not possible to get the required BD and accommodate the cavity and mould size on 1000T Press. Edge pressing may be considered if daylight and configuration are suitable.

In this case, 375 will be pressing height and pressing surface area will be 500 X 100 = 500 cm^2

No of cavities with Specific pressure of 0.8 tonne / cm^2 will be 1250 / 500 = 2.5. Two cavities can be designed.

The length of brick will be along the width of mother mould and width of brick along length of mother mould.

The required cavity length = (width of brick * No of cavity) + (Thickness of liner at top of mould cavity*No of cavities *2) + Thickness of middle Packing + (End packing thickness* 2)

= (100 * 2) + (29 * 2*2) + 14 + (35 *2) = 400 mm

The required cavity width = Length of brick + (liner thickness at top *2) + (Packing thickness*2)

= 500 + (29 *2) + (20*2) = 598 mm

Wall thickness in mother mould = 910 – 598 = 312 / 2 = 156 in **width side**

Wall thickness in mother mould will be 1050 – 400 = 650 /2 = 325 mm in **length side**

Specific Pressure available with 1000 cm^2 pressing surface area = 1000*1000 / 1000 = 1000 kg / cm^2.

Required cross section of mother mould in **width side** =0.01* 59.8 *√1000 = 0.01* 59.8 * 31.6 = 189.0 mm, where as it physically 156 mm.

The required wall thickness of mother mould in **length side** = 0.01 X 40.0X √1000 = 126.4 mm, whereas available thickness is 323 mm.

The laying of brick length, width wise in mother mould is not possible as mould may crack. The length of brick has to be laid along length of mother mould

In this case Length wise cross section will be 1050 - 598 = 452/2 = 226 mm and widthwise cross section will be = 910 – 400 = 510 / 2 = 255 mm.

In this provision, the mother mould is capable of observing pressing load but mixture charging and brick lifting facilities are to be studied and explored for implementation.

 Let us consider designing of mould for 300 X 150 X 100

Possibility of moulding 300 X 150 X 100 brick by flat pressing system

In this case pressing surface area will be 300 X 150 = 450 cm^2

Maximum cavity size in centre of mother mould with 0.8 tonne / cm^2 specific pressure

= 1000 / 0.8 = 1250 cm^2

No of mould cavities for this brick will be 1250 / 450 = 2.77

Specific pressure available with 2 cavities = 1000 / 900 = 1.1 Tons / cm^2

Length wise cavity required in mother mould for two cavities of bricks = 150*2 +29*2*2 + 25*2+ 14 = 480 mm

Length wise wall thickness in mother mould = 1050 – 480 = 570 / 2 = 285 mm

Width wise cavity size in mother mould for two cavities of bricks = 300 + 29*2 +25*2 = 408 mm

Width wise wall thickness in mother mould = 910 – 408= 502 / 2 = 251 mm

Cavity size in the centre of mother mould to accommodate 29 mm thick liners, 20 mm thick pacing plates and 14 mm middle packing will be = 480 * 408

The height of mother mould = brick thickness + 80 % of brick thickness + bottom die thickness = 100 + 80 + 50 = 230 mm

Finally mother moulds with an outer size of 1050 X 910 X 230 have 480 * 408 cavity size in the centre. This exercise and calculations are based on an individual basis. The grouping of other sizes of bricks is to be considered for economic considerations. The height of mother mould can't be taken as 230 mm as pockets and other feature will get disturbed. It must be 250 mm.

Tap holes design:

 The design of tap holes on top and bottom surfaces of mother mould is very important for joining with lock plates. There is not any hard and fast rule for position and size of tap holes but consideration of following points are very important.

- The tap holes size and length should be strong enough to hold the liners and packing rigidly in position with mother mould.
- The assembly should wear the compressive load of press on brick mixture during operation.
- The tap holes centre should not be far away from the end point of the cavity. It should be 25 mm away from the end of cavity edge or centre to centre length should be = cavity length +50 mm for 1000 tonnes press. The tap size and centre distance may change for the higher capacity press. It is applicable for width side also.
- There will be a chance of crack if the centre is very near to end.
- The lock plate may bend if it is far away from the end of the cavity.
- The tap holes centre distance should be at equal space along the four sides of mother mould cavity.

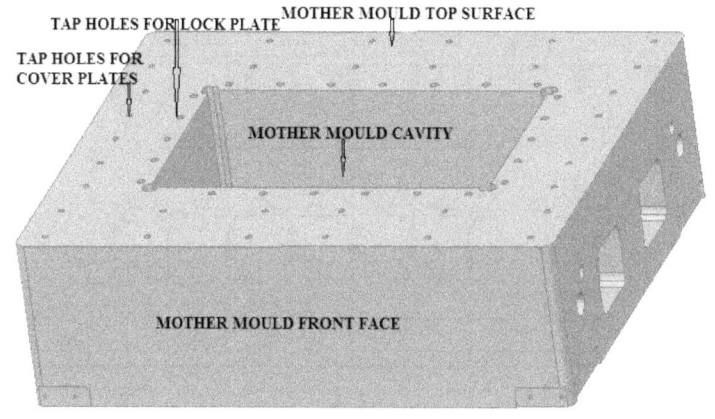

TAP HOLES FOR LOCK PLATE

MOTHER MOULD TOP SURFACE

TAP HOLES FOR COVER PLATES

MOTHER MOULD CAVITY

MOTHER MOULD FRONT FACE

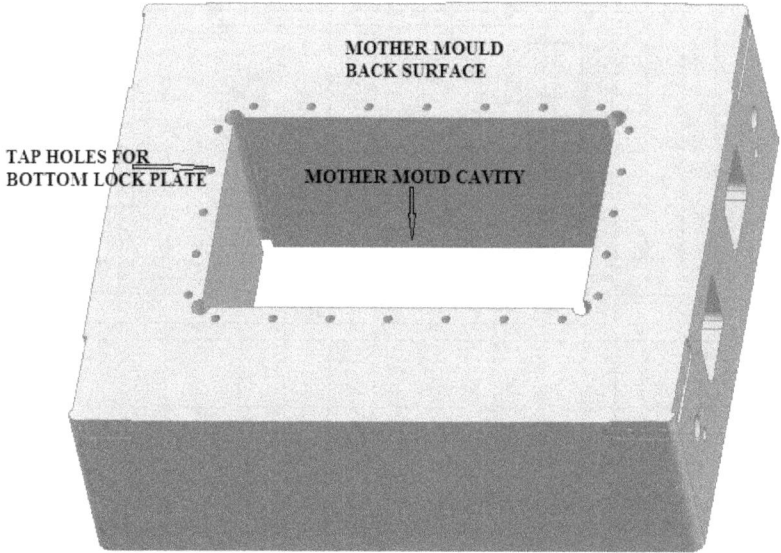

MOTHER MOULD BACK SURFACE

TAP HOLES FOR BOTTOM LOCK PLATE

MOTHER MOULD CAVITY

Tap holes beyond lock plate area on top face:

These are smaller size of tap holes than for holding lock plates. It is used for holding cover plates. All the tap holes centres should be equally distributed. It should be strong enough to hold the cover plates in position. The size and depth have been given in drawing. It may vary according to the requirement as justified by the designer for the higher capacity press.

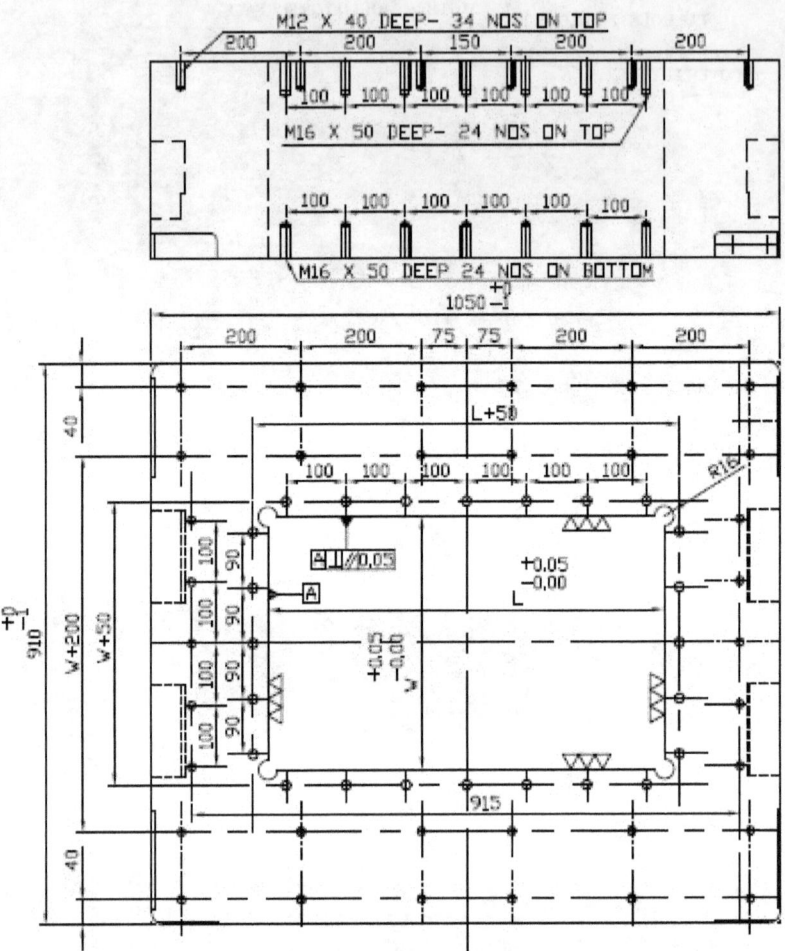

M16 X 50 deep – 24 Nos on top and bottom face given in above drawing are for holding lock plates. M12 X 40 deep – 34 Nos at top surface in above drawing is for fixing cover plates.

Chapter – 7

Lock plates and its application

Design of Lock plates:

The lock plates are very important part of the mould assembly. It maintains rigidity in the assembly of liners and prevents downward and upward movement during compressive force on mixture and ejection of bricks. The lock plates can be designed to individual requirement of the plant. It is designed to

1. Support full thickness of liner: In this case liners will be under top and bottom lock plates. The height of liners will be equal to the height of mother mould.
2. Support half thickness of liners at both ends: In this case liner height will be equal to the height of mother mould + total thickness of top and bottom lock plates. This design is made to adjust the filling depth when mother mould height is less than requirement.
3. Support half thickness of liner at the top and full thickness at the bottom: In this case liners thickness will be equal to the thickness of mother mould + thickness of top lock plate.

Organs of lock plate:

- Mainframe body made with suitable grade of steel
- The cavity in the centre of the main body: The cavity size must permit, clear up and down movement of dies with a plunger. The cavity boundary lines should match with chamfer line of liners top and bottom face.
- Counterbore drill holes to attach with Allen bolts on top and bottom surface of mother mould
- Counterbore drill holes to attach with packing plates if needed.

Mainframe body: It is made with a suitable grade of steel mostly C – 45.

Cavity Size: It is generally made 4 – 6 mm bigger than mould cavity size for clear movement of top and bottom dies.

Counterbore drill holes: There are two different sizes of counter drill holes. The bigger size is for fixing with mother mould. The smaller size is for fixing with side and end packing. The drill holes size should be 2 mm bigger than the size of thread and body of Allen bolt. The counter drill size also should be 2 mm bigger than the head of Allen bolt which is generally 1.5 times the size of Allen bolt.

The different systems of lock plate design:

Three different systems of lock plates are represented here in 3D drawings.

Liners having undercut at the top surface: The end liner has undercut at the back side and the side-liners have undercut at both ends from the top surface. It has been drawn and marked in 3D drawings. This system is used to reduce mould height and adjust with the daylight of press. The bottom lock plate will sit directly on the bottom face of liners assembly.

A liner having an undercut at both sides (Top and Bottom face): The bottom face also has the same design of liners and lock plate. The liners height will further increase equal to the thickness of bottom lock plate.

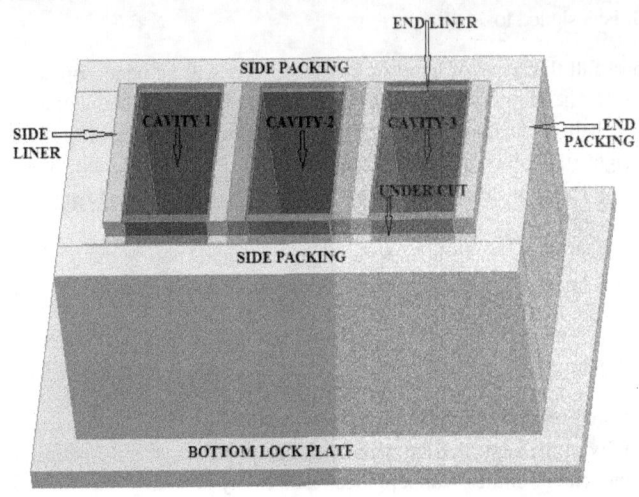

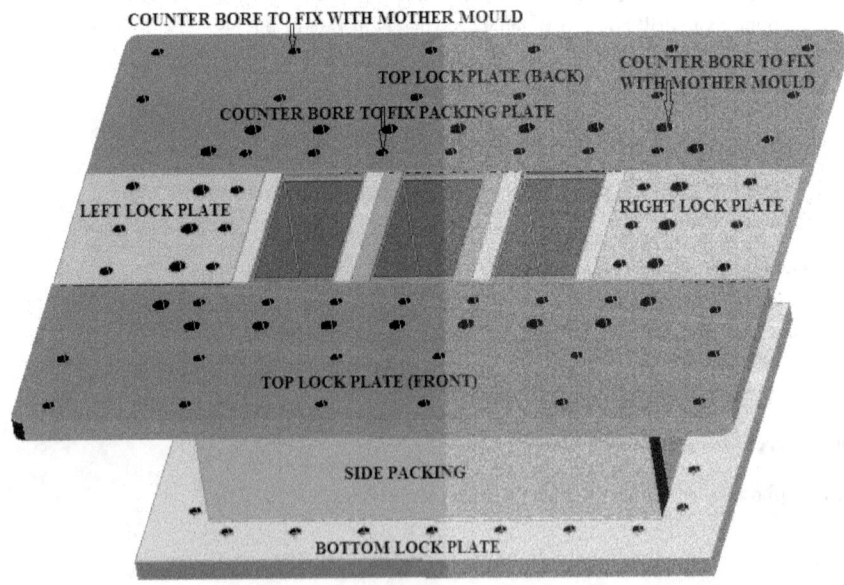

The front and back top lock plates represented in the 3D drawing are the combination of cover plate and lock plates.

Liners having top and bottom face plane surface: In this case, there is no undercut. The lock plates will sit on the plane surface. The drawings given here are self-explanatory. It is done to reduce extra work and increase the strength of liners at both ends whenever possible to design.

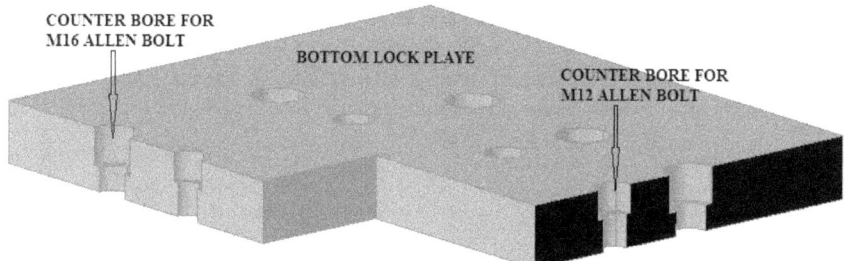

SHAPE OF COUNTER BORE DEPICTED IN SECTIONAL VIEW

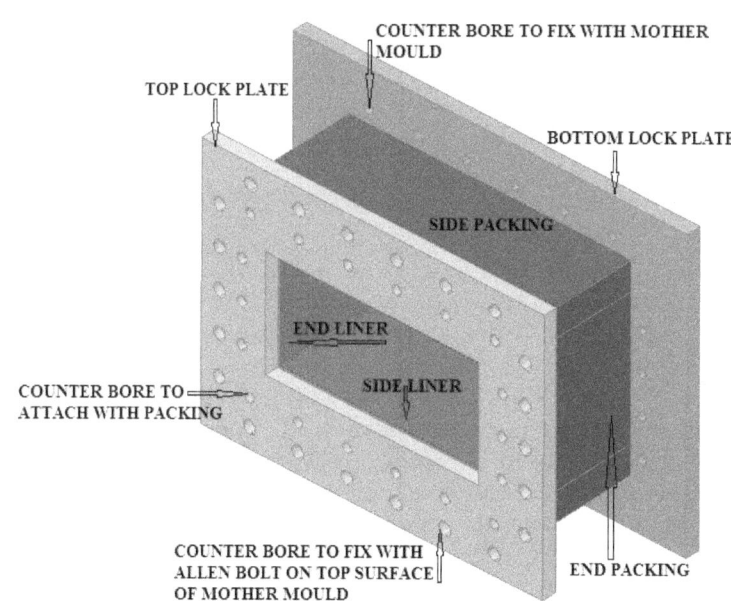

Chapter – 8

Liner, Loose piece and packing design

Design of side-liners and end liners:

The liners form the peripheral boundary of the brick shape. It forms the negative shape of bricks bounded by four sides in case of plane shape brick. The liners are exposed to pressing and ejection force on each cycle of operation. The bricks are to be ejected smoothly without any internal or external cracks or damage. It also has to resist deformation and maintain rigidity during operation. The face tapers are provided on liners surface that comes in contact with the brick surface to get bricks within tolerance limits.

The design of liners depends on

- Placement of mould cavity for production of bricks in mother mould
- Shape and size of brick
- Calculated number of cavities on the basis of pressing capacity and specific pressure.

Tolerance:

The tolerance is a permissible deviation from basic nominal sizes of bricks. Generally, bilateral tolerance is given on basic nominal size such as ± 1.0. It means bricks will be acceptable 1 mm bigger or 1 mm smaller than the basic size. The permissible taper is also provided by customers.

Design considerations:

Side liners

- Select the composition of liner material to have required hardness.
- Decide hardness of liners to increase life.
- The length of side liner should be the length of brick + twice the thickness of end liner.
- The height of side liner (width) should be equal to the height of mother mould. The thickness of lock plate may be included or excluded according to design criteria.
- The side-liners should have a draft taper on the face that will be in contact with the brick surface. The face taper is to be decided and maintained according to required taper in bricks.

Organs of side liner:

- Body: It is made of High Carbon high chromium steel or any other suitable material that can gain abrasion resistance after heat treatment.
- Chamfers at two inner edge on top and bottom faces: General size is 3X3 mm
- Face taper: It is a common feature for all liners provided on the inner face of mould cavity walls coming in contact to brick surface.

- Step down and undercut: It is provided when liners have to expose above mould height on both sides top and bottom or only on the top side.

Design of face taper:

The face taper is designed on the basis of permitted tolerance on brick and taper on four sides of brick that will come in contact with liners. Providing 1 mm face taper on liners will serve the purpose but in the certain case where the specification is very stringent and no mortar is applied in between bricks taper should be plotted as in given drawing.

Exercise:

Let us consider a brick of size 300 X 150 X 100

$$\text{Permitted tolerance} = \begin{matrix} -0.5 \\ -1.0 \end{matrix}$$

Permitted taper 0.5 mm from top to bottom

In this condition, brick size can be 300.5 maximum at the top and 299.0 minimum at the bottom.

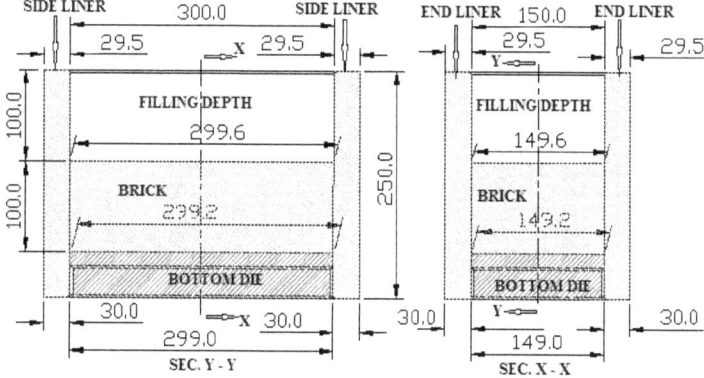

Fix the brick size at the top and bottom face to have the required dimensional tolerance and taper. Reduce 1.0 mm from basic nominal size from length and width. You will get 299 x 149 by doing it. Let us consider 299.2 X 149.2 which is slightly bigger than minimum required size. The permissible taper is 0.5 mm. Let us consider it 0.4 mm which is slightly less than permissible taper. The brick size at top face will be 299.6 X 149.6 mm with 100 mm. the thickness of the brick. Plot the brick size in width also as in above drawing; extent both end's vertical lines to a distance of 100 mm on top and 50mm on a bottom side. Take 30 mm liner thickness at the bottom on sides of brick and extend it to 250 mm upwards. Follow the same procedure for all bricks where such tolerance and taper are provided.

End liners

- Select composition of liner material to have required hardness.
- Decide hardness of liners to increase life.
- The width of end liner should be equal to the width of brick + draft taper at both ends to match with a draft of side liners. The surface of end liners towards cavity must have draft taper matching with taper required in brick ends. It should also have a matching profile, shape or angle of brick's end face.
- The height of end liners will be equal to the height of side liners.
- Permissible draft in brick is also considered in deciding face taper in the end liners.

Follow the procedure to have the width, height and thickness with the draft as plotted above.

Design category: The shapes of liners are given under each category. Visit lock plate chapter to see placement of lock plates

Category - 1: In this system, undercuts is provided as marked in the drawing. It facilitates seat to top lock plate that holds rigidly in position with bolts on the face of mother mould. The height of liner is equal to the height of mother mould + thickness of top lock plate.

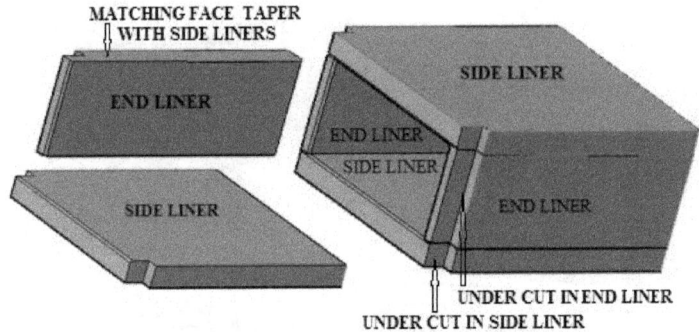

Category – 2: In this category, under-cuts are provided on bottom side also as in top. It provides a seat for bottom lock plate also as in top. The height is equal to mould height + top and bottom lock plate thickness.

Category – 3: In this category top and the bottom face of liners have plane surface without any undercut. Top and bottom lock plates are placed on these surfaces up to the chamfered edge.

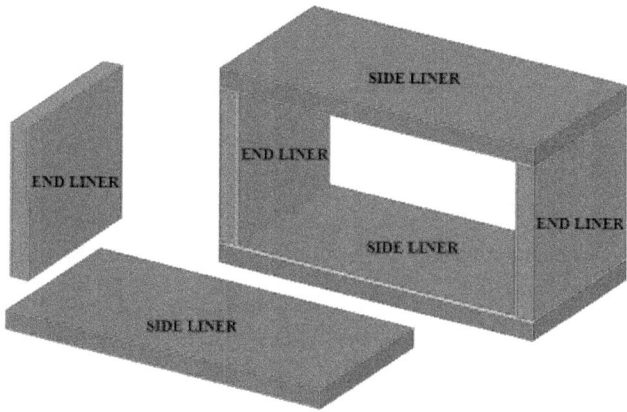

Loose pieces:

The loose pieces are designed to comply with special features of the bricks. It may be on the side, ends, the top or the bottom face of the brick. When features are on sides and end face of the brick, the loose piece is designed and attached with side and end liners to have the required features of a brick. The feature of the top and the bottom face of brick are attached with dies. These pieces are made separately and attached with Allen bolts. The length of these pieces will depend on its position in the cavity and dies. It also depends on size, shape and position of a profile in bricks. The material and hardness will be same as liners and dies. It has to be set suitably in the respective position. These loose pieces will be the negative shape of brick's features.

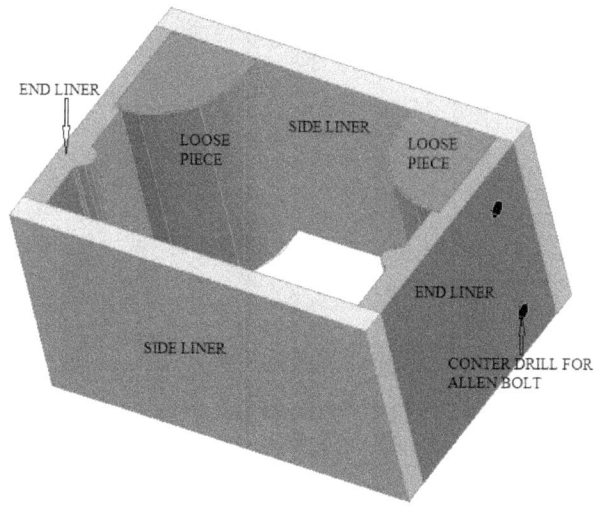

The loose piece in the above figure has tap holes of matching size with Allen bolt and counter drill on end liner.

Packing plates:

The packing plates are provided on the back side of liners to maintain rigidity in mother mould cavity. It is also essential to get the required size in mould cavity within permissible tolerances.

Importance of the packing plates:

- Maintains rigidity in mother moulds cavity during pressing and ejection load.
- It will provide scope to assemble liners for making another size / shape of brick by changing thickness or adding another packing.
- It reduces shock load of horizontal components on mother moulds by observing shearing load to a certain extent.

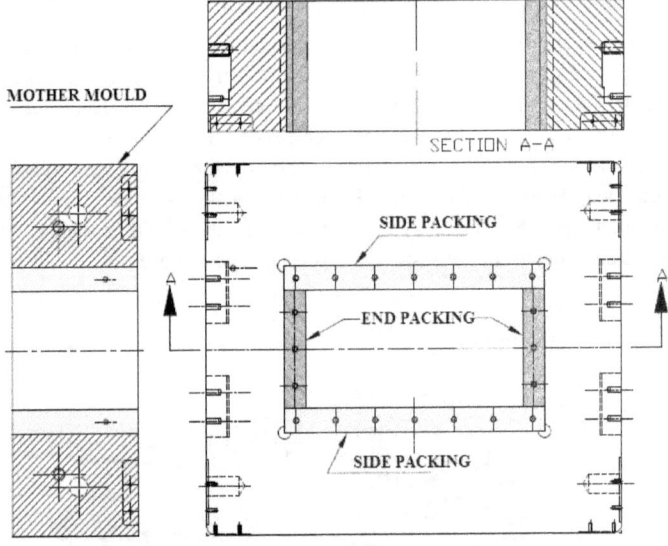

Calculation of packing sizes in fixed cavity sizes:

When length of brick is along width of mother mould:

End Packing thickness = [Length of cavity - {(Brick width at the bottom of mould cavity * No of cavities) + {(Thickness of middle packing * (No of cavities - 1)} + (Thickness of liner at the bottom side of cavity * No of cavities * 2)}] / 2

Side Packing thickness = Width of cavity in mother mould – (Brick length at the bottom of mould cavity + Liner thickness at bottom side * 2) / 2

The thicknesses of middle packing plates depend upon the gap between two arms of the gripper and its expanded width to hold the bricks.

When length of brick is along length side of mother mould:

Side Packing thickness = [Width of cavity - {(Brick width at the bottom of mould cavity * No of cavities) + {(Thickness of middle packing * (No of cavities - 1)} + (Thickness of liner at the bottom side of cavity * No of cavities * 2)}] / 2

End Packing thickness = Length of cavity in mother mould – (Brick length at the bottom of mould cavity + Liner thickness at bottom side * 2) / 2

Packing seems:

The surface of liners worn out due to successive use for prolongs period. It is practice to regrind the surface of lines that come in contact of the abrasive grain of mixture and reassemble for further use. The liners thickness will reduce after grinding. The required thickness of seems packing is fitted in between packing and liners or back side of packing to maintain rigidity and previous position of the mould cavity.

Chapter – 9

Plunger design

Top and Bottom Plungers:

The designs of top and bottom plungers are very important for the formation of brick shape. The lengths of plungers are decided on the basis of operating system of press and configuration. The cross-section (end face) is designed to match with the geometry of brick shape (peripheral boundary of mould cavity). You will get different shapes of plungers with pockets in the exercises.

Organs of Plungers:

- Length
- Upper-end face
- Lower end face
- Side faces
- Pockets
- Drill holes

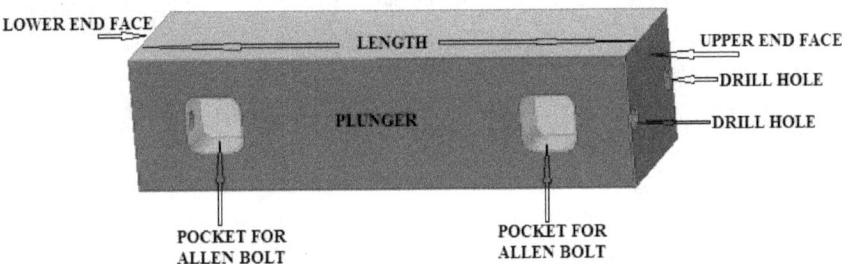

Length: The length of top and bottom plungers are decided

- To suit the operating parameters and press daylight.
- To have sufficient strength in cross-sectional area to resist pressing load of press while mixture is being compressed in between two dies

Upper and Lower End Face: There are two end faces, one at the top and another at the bottom of plunger length. The length and width of end face should be about 2-3 mm less than opposite face of the die's pressing surface. Both faces should be a smooth, parallel and right angle to each side of length individually within specified tolerance limits. Each face must have at least two drill holes to insert Allen bolts through pockets for attaching dies and plunger holding-plate. The bottom end face of the bottom plunger is attached with plunger holding plates and top face with the bottom die. Similarly top face of top plunger attached with top plunger holding plate and bottom face attached to the top die. **There should be one tap hole in the centre at top and bottom face for handling purpose.**

Pockets: The shape, size and location of pocket in plunger are very important. It must be designed considering following points

- Place the pocket on opposite sides of the thicker cross section.
- The pocket should not reduce the strength of plunger that has to wear the pressing load of press.
- The length of pocket should be large enough to insert Allen bolt larger than the length of the drill hole.
- The width of pocket should be wide enough for free movement of Allen key for tightening with tapped holes in dies and plunger holding plates.
- Depth should not make plunger weak but it must provide clear movement to Allen bolt head while fitting with dies and plunger holding plates.
- The metal thickness above pocket should be adequate to fit bolt and take the load.
- The no of pockets and location decided according to the geometrical shape of the end face and cross-sectional area.

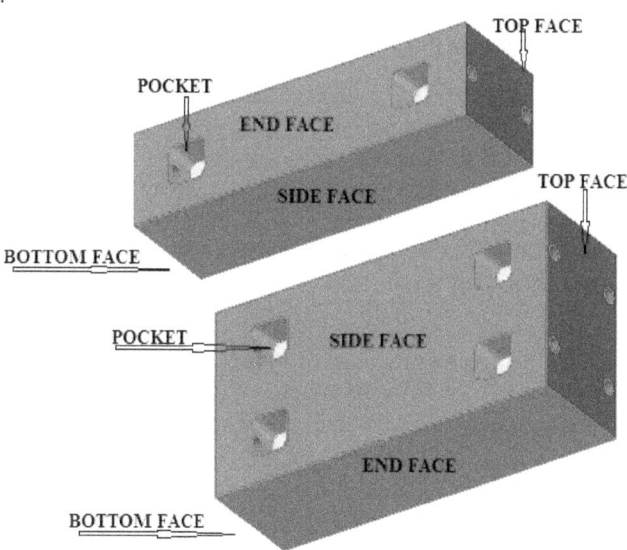

The end and side face here have been marked considering length and width of top and bottom surface of the plunger to indicate numbers and position of pockets. The length here does not relate to the length of plungers.

Drill hole: Position and length of the drill hole are very important.

- It should be located at an appropriate distance from the side of the plunger. The hole should not shear off from the side while fitting bolt. It will be clear in 2D plunger drawing
- The length should be 40 -50 mm depending upon size and length of the bolt to be fitted

Operating systems of press and plunger length calculation:

The plungers are designed to suit prevailing operating systems of a hydraulic press. Go through each operating system and follow the equation to calculate plunger length

The operating systems available in press are

1. Fixed bottom , floating mould table and top pressing cylinder
2. Top and bottom pressing system.
3. Fixed top, the bottom pressing cylinder and floating mould table.
4. Fixed mould table, top pressing cylinder and ejection cylinder from the bottom.
5. Top and bottom pressing cylinder with floating mould table and pin cylinder from bottom

Fixed bottom with floating mould table and top pressing cylinder:

In this system, the bottom plunger with die remains fixed. The mould moves up and the required filling depth is maintained by setting the upward movement. The mixture is filled into the cavity by the charger in this position. When charger moves back, the top pressing cylinder moves down with the top plunger and top die to press the mixture in the cavity. The top pressing cylinder continues to press the mixture as per setting till brick's required thickness is attained. Daring is also set to remove entrapped air and for better setting of refractories grains.

The top pressing cylinder moves up after pressing the brick to required thickness. The mould table with mould moves downward after pressing sequence is over. The brick remains stationary on bottom die. In this process, the brick is ejected out of mould cavity. The gripper is being set at the appropriate height to hold the brick properly for removing it from die.

Design of top plunger:

The shape and size of top plunger are governed by

- The top geometrical shape of a brick and its surface area.
- The total filling depth and brick thickness.
- The stroke length of top pressing cylinder
- The stroke length of mould lifting table's cylinder.
- The gap between brick takes off height and top pressing cylinder.
- Height of charger
- Height of gripper

The length of the top plunger is very important. It connects top ram of pressing cylinder and die. A plunger holding plate 50 mm thick is provided in between top ram and plunger to connect each other. A top die is generally 50 mm thick is attached at another end of the plunger.

The length of Top Plunger = daylight – (height of bottom plunger with the holding plate and die + stroke length of top pressing cylinder +30) mm.

The length of the top plunger should not be very high. The long plunger with comparatively thin cross section will bend and bulge out. In such condition spacer with the higher cross section is provided to reduce the top plunger length. Study corrections in the daylight of press configuration for making 50 to 200 mm thick bricks.

In this case, top plunger length with plunger holding plate and die = Distance from bottom die to bottom face of spacer – (stroke length of top pressing cylinder + 30)

30 mm has been subtracted just to keep the distance more than stroke length of top ram. The brick pressing thickness should be more than 30 mm. If you want to press brick less than 30 mm you increase space or plunger length accordingly.

Design of bottom plunger:

There is no movement of the bottom plunger in the above operating system. Its function is to resist the pressing load of the press while compressing brick mixture to have required size. The brick is also ejected by it from mould cavity when mould moves down. The shape and size of bottom plungers are governed by the

- The bottom geometrical shape of a brick and its surface area.
- Comfortable brick takes off height. – It comprises of
    1. the bottom dies thickness,
    2. bottom plunger height (length),
    3. the thickness of bottom plunger holding plate
    4. the height of bottom plunger fitting base from ground level.
- The gap between the top dies face and bottom die face after fitting on respective positions. - It should be equal to the sum of total filling depth + charger height + clearance.
- Stroke length of mould table.
- Thickness of brick
- Mould height to have required filling depth
- The distance between the top surfaces of the bottom die and mould at the extreme bottom down-level of the sliding table.

Bottom plunger length with a plunger holding plate and bottom die thickness = height of bottom plunger holding base from ground level + height of mould table top at extreme down level from plunger holding-base + height of mother mould + exposed distance of bottom die top surface from mother mould top when in extreme down level.

This total distance from ground level must be at comfortable brick take off level. When it will be at uncomfortable level due to a higher pressing thickness of brick, the arrangement has to be made for taking off brick comfortably.

In the corrected press configuration the comfortable brick takes off height has been taken as 1068 mm. The height of mould table top from plunger holding base is 185 mm for making 200 mm thick brick. This height will increase or decrease with a change in mould height.

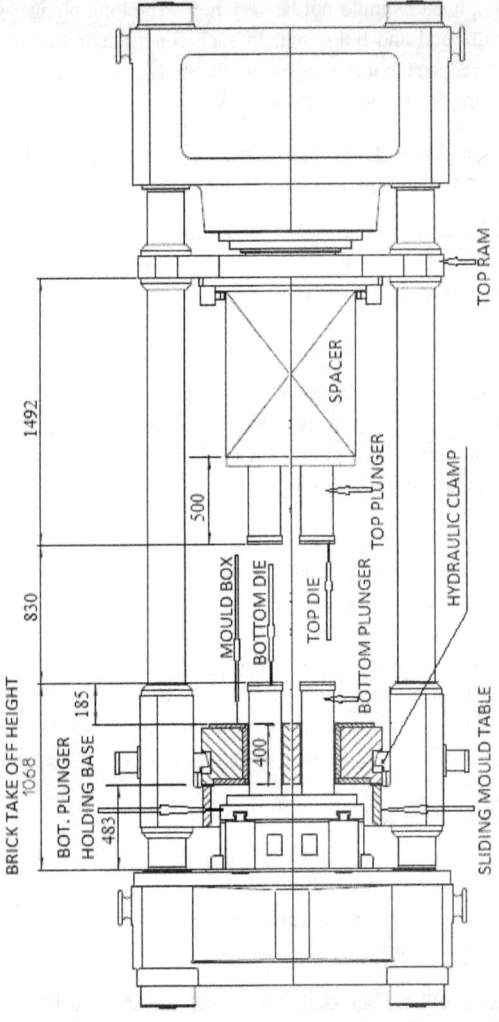

The height of plunger holding base = 368 mm, the height of sliding mould table top = 115 mm, Total height = 368 +115 = 483 mm. The stroke length of top cylinder = 800 mm and stroke length of sliding table = 1250 mm. The press has been configured with detail size of top and bottom plungers.

Top and bottom pressing system:

In this system pressing of the brick mixture is being done by both sides (top and Bottom). The mould table may be fixed or floating depending upon press design and operating system. The bottom pressing cylinder has a dual function. It works as pressing cylinder as well as ejection cylinder. The brick may be also ejected keeping bottom plunger stationary and moving mould up or down in case of floating table as and when required to eject abnormal shape.

Design of Top plunger:

The shape and size of top plunger are governed by

- The top geometrical shape of a brick and its surface area.
- The stroke length of top pressing cylinder.
- The gap between the top pressing cylinder and mould fixing table.
- The maximum stroke length of mould lifting cylinder.
- The stroke length of bottom pressing cylinder
- The size of a brick and filling depth.

The length of Top Plunger = Distance from the bottom die top to Top plunger holding base at mixture charging level – (stroke length of top pressing cylinder + 30). The Gap between mould top surface and top die surface must be more than brick thickness. The gap also should be large enough for free movement of charger and gripper individually.

Bottom plunger design:

Its shape and size are governed by the

- The bottom geometrical shape of a brick and its surface area.
- The height of mould table and bottom pressing cylinder distance
- The stroke length of bottom pressing cylinder.
- Distance between top die and bottom die.
- Pressing thickness of brick and fill depth.
- Brick take off height

Bottom plunger length with plunger holding plate and bottom die thickness = Height of mould table top surface at extreme down level from plunger holding base + 50 mm

Fixed mould table with top pressing cylinder and ejection cylinder from bottom:

In this system, the mould is fixed on mould table. The mixture is filled into the cavity by charger or manually. The top plunger with top die comes down to press the mixture to have required thickness. The top plunger with dies goes back after pressing. The brick is being ejected by ejection cylinder from the bottom.

Design of Top plunger:

The shape and size of top plunger are decided by the

- The top geometry of brick shape and surface area.
- Pressing thickness of brick and filling depth.
- The stroke length of top pressing cylinder.
- The gap between mould table and top ram of the press.
- The height of charger.
- Clearance above charger.

The length of Top Plunger = Distance from the bottom die top to Top plunger holding base – (stroke length of top pressing cylinder + 30) mm.

The Gap between mould top surface and top die surface must be more than brick thickness. The gap also should be large enough for free movement of charger and gripper individually.

Design of bottom plunger:

The shape and size of bottom plunger are decided by the

- The bottom geometrical shape of a brick and its surface area.
- The gap between mould table top and plunger fixing base.
- Ejection stroke length.

Bottom plunger length with plunger holding plate and bottom die thickness = Height of mould table top surface from plunger holding base + 50 mm

Top and bottom pressing cylinder, pin cylinder and floating table:

In this system, the pressure is applied from top and bottom to compress the brick mixture to required size of a brick. The function of pin cylinder is to attach the pins in case of hollow bricks such as nozzle and checker bricks. The table may remain stationary or move up and down depending upon required operational sequence.

The length and size of plungers are calculated as explained earlier.

This press has some special function that is useful for making the special shape. It will be elaborated in next volume.

Chapter – 10

Press configuration and related calculations

The working parameters of a particular press have been given here as an example to explain the process of calculating the maximum brick size that can be made. The press configuration also can be changed to cater the regular requirement of pressing brick thickness.

Working parameters of press:

Daylight= 3390 mm

Stroke length of top Ram =800 mm

Stroke length of sliding table cylinder = 1250 mm

Charger height = 340 mm, clearance above charger = 40 mm (minimum)

Gripper height = 160 mm

Plunger holding base from bottom = 368 mm

Mould table top from bottom = 483 mm

Calculation of maximum brick size that can be pressed:

Let "X" is the brick thickness that can be made on the press with above configuration.

Total filling depth (height) in mould cavity = $X + 0.8*X = 1.8X$

Mould height = $1.8X+50$, Where 50 mm is the die thickness that will be inside the mould cavity.

Bottom Plunger height with plunger holding plate and bottom die = Height of sliding mould table from ground level + Mixture filling depth in mould cavity + Bottom die thickness + Height of die above mould face at extreme down level.= $483 + 1.8X + 50 + 20 = 553 + 1.8X$

Top Plunger height with plunger holding plate and top die = Press daylight – (Bottom plunger height with plunger holding plate and bottom die + Brick thickness – 20 + Stroke length of top ram) = $3390 – (553 + 1.8X + X – 20 + 800) = 3390 – 1333 – 2.8X = 2057 – 2.8X$

20 mm has been subtracted to allow the movement of top plunger with die below brick thickness that can be restricted by control panel.

Sum of top and bottom plunger height with plunger holding plates and dies = $2057 – 2.8X + 553 + 1.8X = 2610 – X$

Gap between bottom and top dies = Press daylight – Sum of top and bottom plunger height with plunger holding plates and dies from respective base = $3390 – (2610 – X)$

$= 780 + X$ ----- (1)

Gap between two dies considering charger height, clearance above charger and total filling depth in mould cavity = Charger height + Clearance above charger + Total filling depth in mould cavity = 340 + 40 + 1.8X = 380 + 1.8X ------- (2)

Clearance above charger is variable but to be decided for free forward and backward movement of charger.

(1) Must be equal to (2)

780 + X = 380 + 1.8X

Or 780 – 380 = 1.8X – X

Or 400 = .8X

Then X = 400/0.8 = 500 mm

It concludes that maximum 500 thick brick can be made on this press.

Filling depth = 1.8 X = 1.8*500 = 900 and mould height = 900 + 50 = 950

Bottom Plunger height with plunger holding plate and bottom die = 553 + 1.8X = 553 + 1.8 *500 = 553+900 = 1453 mm

If thickness of plunger holding plate and bottom die is 50mm each

The bottom plunger length = 1453 – 100 = 1353 mm.

Top plunger length including plunger holding plate and die = 2057 – 2.8X

= 2057 – 2.8*500 = 2057 – 1400 = 657 mm

The press has been configured accordingly to press mixture for making 500 mm thick brick.

The above equation is applicable to any press.

The press has been configured as shown in a sketch with above calculations. The press is showing mixture charging operation with charger above the mould surface. The clearance of 40 mm above charger surface is also marked.

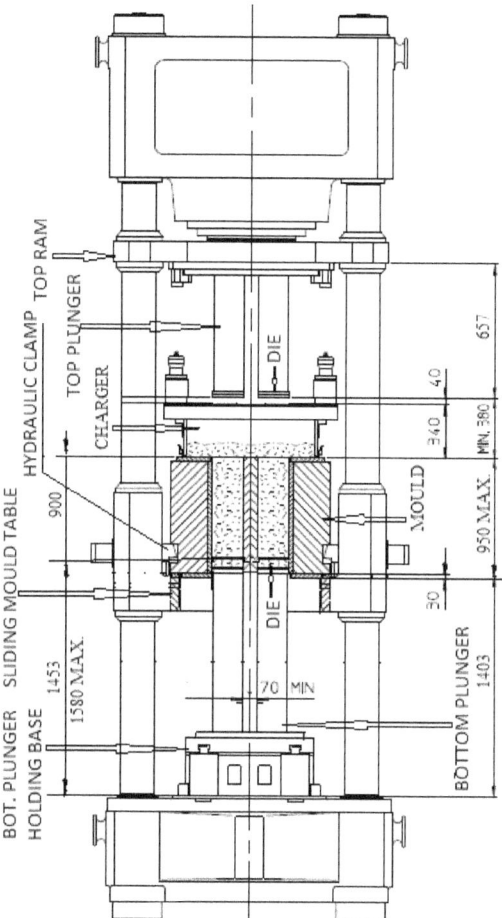

FRONT VIEW OF PRESS SHOWING MIXTURE CHARGING

The mould lifting cylinder has moved up with mould to receive mixture in the above drawing. The mixture has been charged in the cavity. The cavity depth is equal to total filling depth. The charger will go back and the mixture will be pressed with a top die by moving downward with the top plunger to have 500 thick brick.

The sliding mould table can be set to move down to certain length along with the down-ward movement of the top plunger. The top plunger will move down equal to fill depth (400 mm) length. In this process pressing zone will change and provide bottom pressing effect.

The thin cross section of brick having 500 pressing height cannot be made as bottom plunger may bend. The spacer under bottom plunger can be provided considering the downward movement of the mould to make such bricks if needed.

Pressing sequence:

The line diagram is showing the pressing sequence of press. The top ram with plunger holding plate, plunger and die has moved down, compressed the mixture to have 500 mm thick brick.

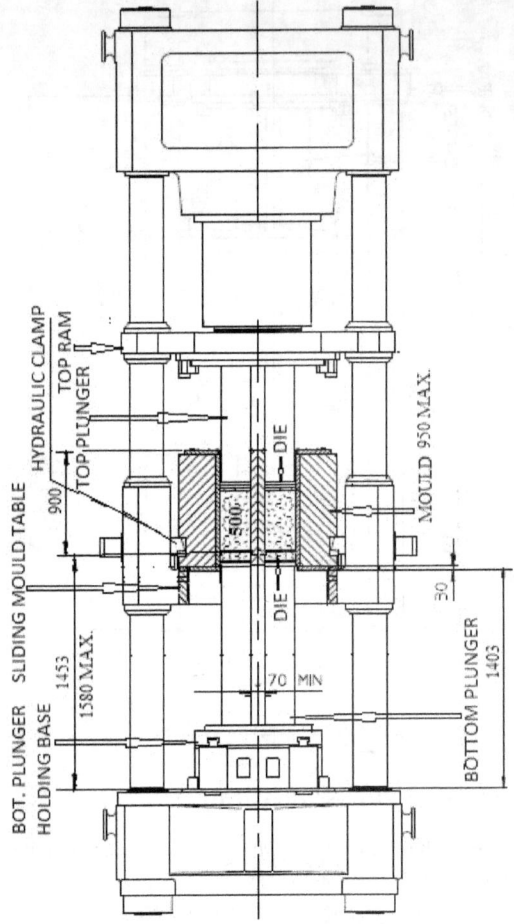

Ejection sequence:

The diagram is showing the ejection sequence of brick. The top ram has moved up along with plunger holding plate, plungers and dies after pressing the bricks. The sliding mould table has moved down leaving the bricks on the top surface of dies. The bricks have been ejected out from mould cavity in this process

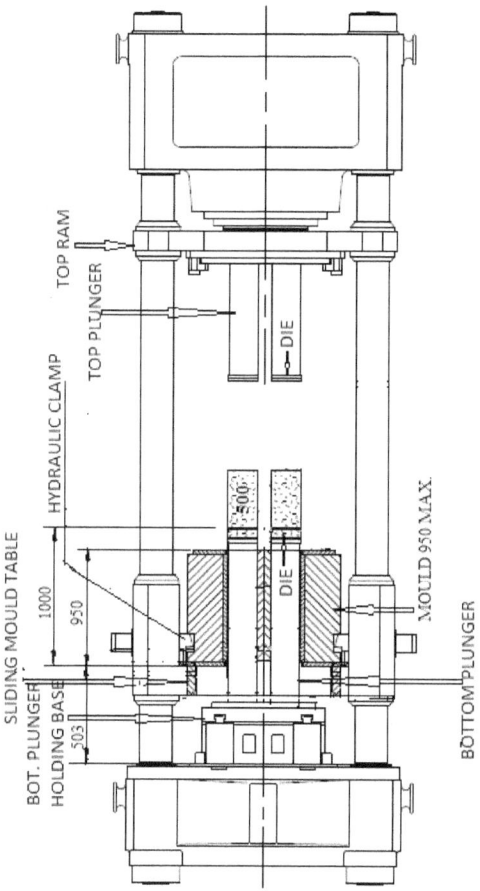

Correction in configuration for making 50 to 200 mm thick bricks

Let "200 mm" is the brick thickness that is to be made in this press

Total filling depth (height) in mould cavity = 200+ 0.8*200 = **360 mm**

Mould height = 360+50, Where 50 mm is the die thickness that will be inside the mould cavity, take it as 400 mm

Bottom Plunger height with plunger holding plate and bottom die = Height of sliding mould table from ground level + Mixture filling depth in mould cavity + Bottom die thickness + Height of die above mould face at extreme down level.= 483 + 360 + 40 + 185 = **1068 mm**

40 mm thickness of die will be inside mould cavity when mould will move up to receive mixture. 185 mm is the extra length added to make brick take off height comfortable.

Top Plunger height with plunger holding plate and top die = Press daylight – (Bottom plunger height with plunger holding plate and bottom die + Stroke length of top ram + brick thickness – 20) = 3390 – (1068 + 800 + 50 – 20) = 3390 – 1898 = **1492**

The tap ram travel distance has been increased by 20 mm for better pressing effect. The top plunger 1492 mm long is liable to bend under pressure with the thin cross section. The change of plunger for moulding bricks of other cross section will not be economical too.

It is better to reduce the length. The plunger length of 400 mm with 50 mm plunger holding plate and 50 mm die thick is appropriate. Fix spacer with spacer holder to comply with the requirement.

The length of spacer with spacer holder = 1492 – 500 =992 mm.

Top plunger height under spacer with plunger holder and die (50 mm thick each) = **500 mm**

Note: The minimum compressed brick thickness has been taken as 50 mm. This adjustment or configuration will work for comprising brick thickness up to 250 mm. The movement of the top plunger with die can be restricted to required pressing thickness by the control panel.

Filling depth in mould cavity =360 mm

Total height of mould top face = 1068 + 360 = 1428 mm

Gap between mould top face and top die = 3390 – (1492 + 1428) = 3390 –2920 = 470

Charger height from mould top face = 340

Clearance above charger = 470 – 340 = 130 mm

Configuration of above press for making 50 to 200 mm thick brick:

Upward movement of mould at mixture filling level = 360 mm.

Charger height = 340 mm.

Gap between mould top face and top die bottom face at charging level = Daylight – (height of bottom plunger with holding plate and die + total filling depth at mixture charging level + length of spacer, top plunger holding plate, plunger and die)

= 3390 – (1068 + 360 + 1492) = 3390 – 2920 = 470

Clearance above charger = 470 – 340 = 130

Gap between two die face attached with plungers = Daylight – (height of bottom plunger holding plate, plunger and die + height of top plunger, holding plate die and spacer) = 3390 – (1068 +1492) = 3390 – 2560 = 830 mm

The height of top die face from ground level = 1068 +830 = 1898 mm

The line diagram of Press has been first drawn with spacer fitted in its position. The cross section of spacer should be large enough to attach plunger holding plates. Next diagram is the configuration of Press equipped with plunger under spacer and a mould for making 200 mm thick brick.

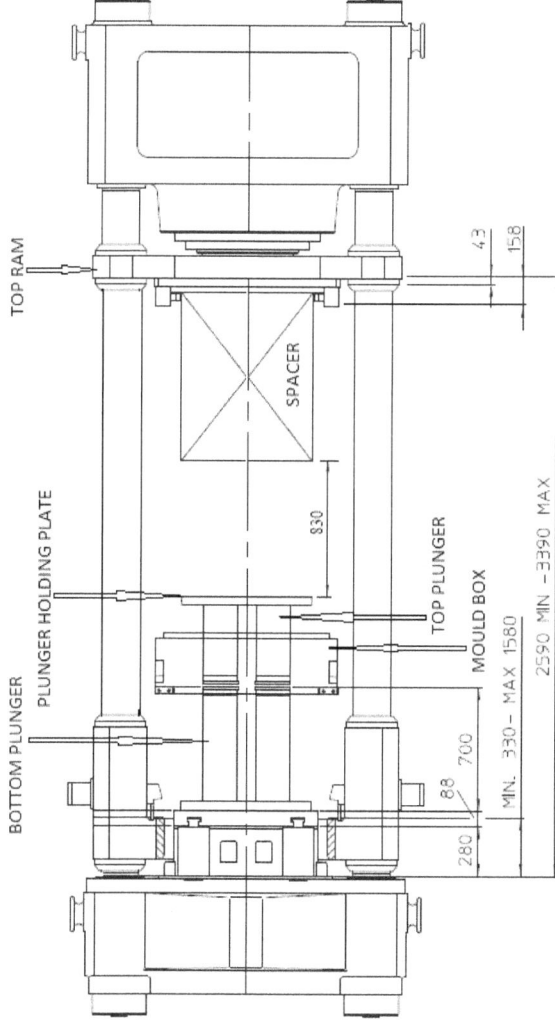

PRESS EQUIPPED WITH SPACER ON TOP RAM

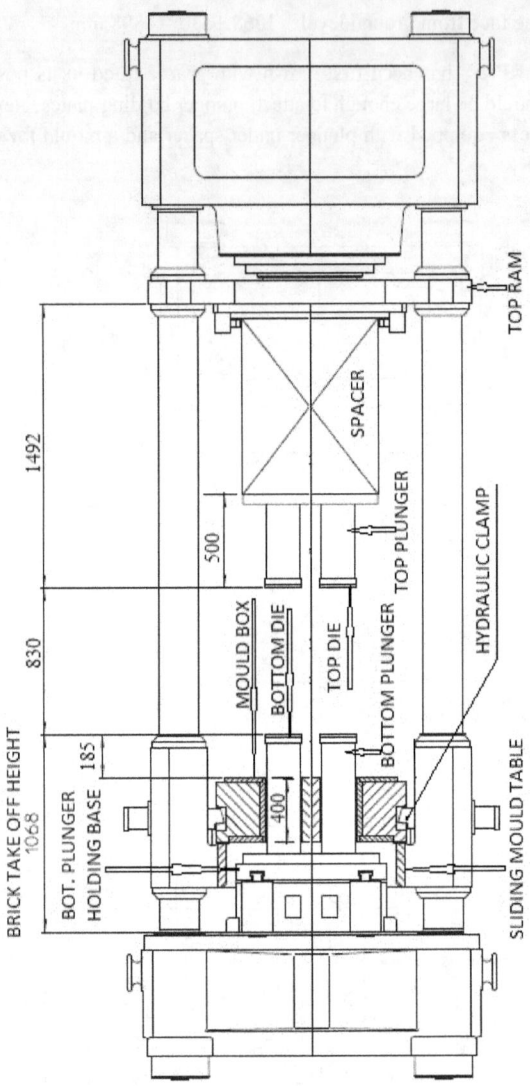

PRESS CONFIGURATION FOR MAKING 200 MM THICK BRICK

Configuration for making 75 mm thick bricks:

The height of top and bottom plunger will be same as described earlier, the only height of bottom die upper face has changed from 185 to 235 mm due to a reduction in mould height.

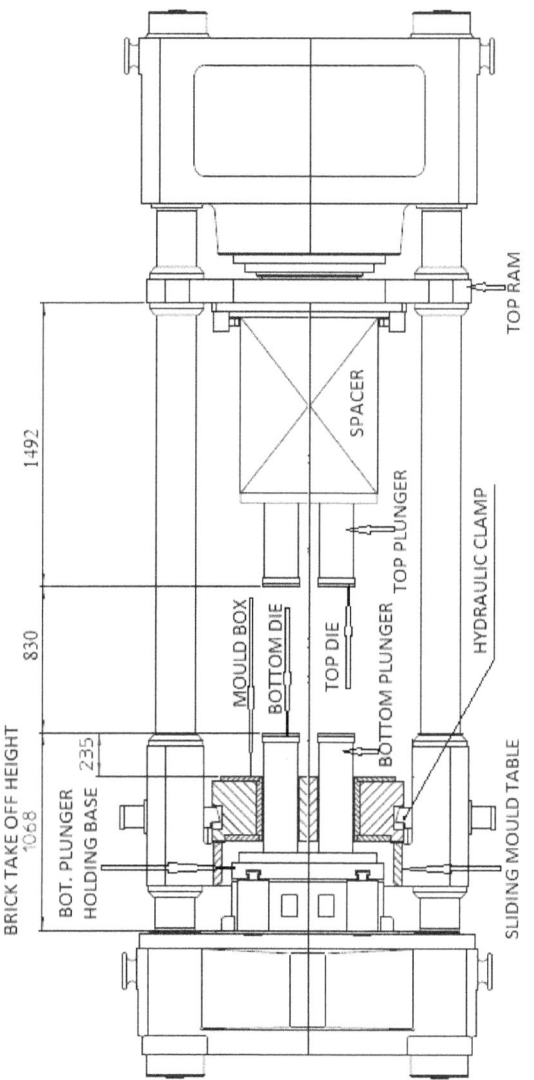

PLUNGERS AND MOULD ATTACHED IN RESPECTIVE PLACE

The sliding table has moved up to receive mixture from the charger.

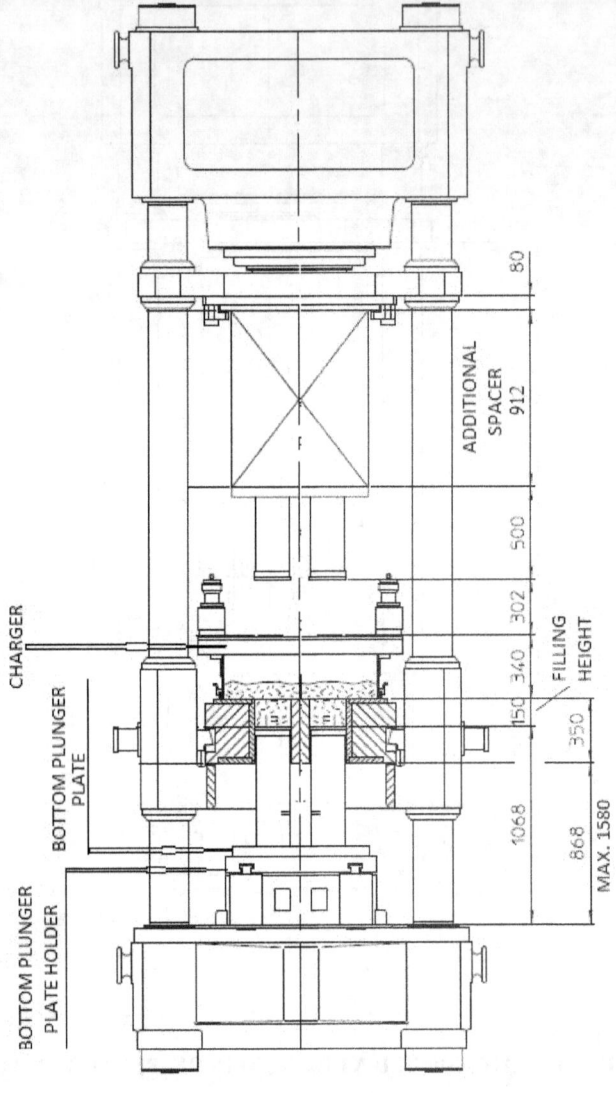

MIXTURE CHARGING IN CAVITY

The charger has gone back after charging the mixture. The top plunger with die came down and pressed the mixture.

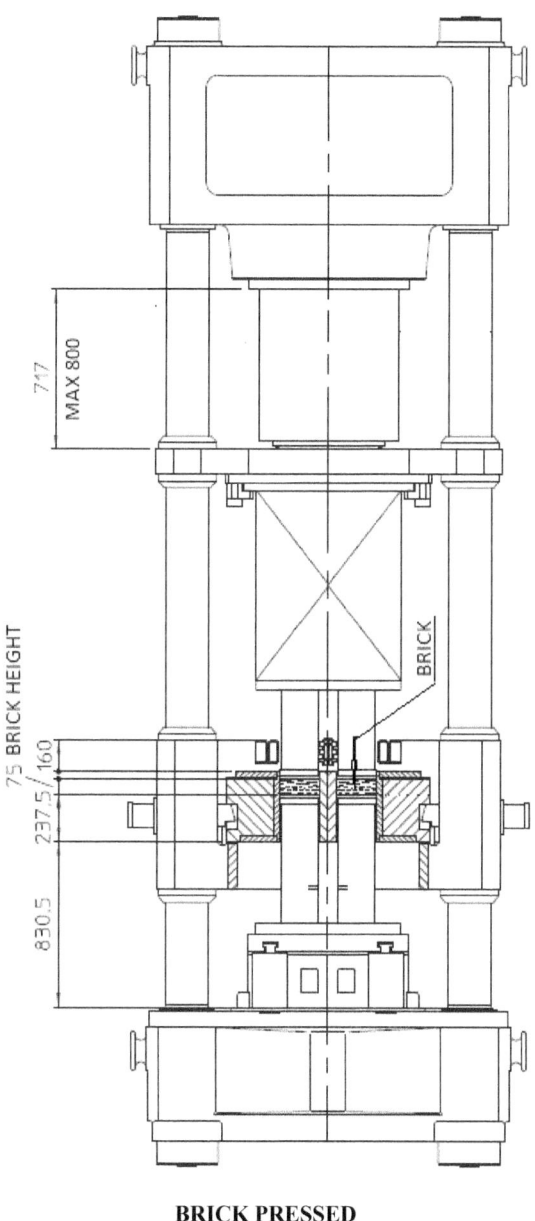

BRICK PRESSED

The brick remained on the bottom die when mould table with mould went back to neutral position.

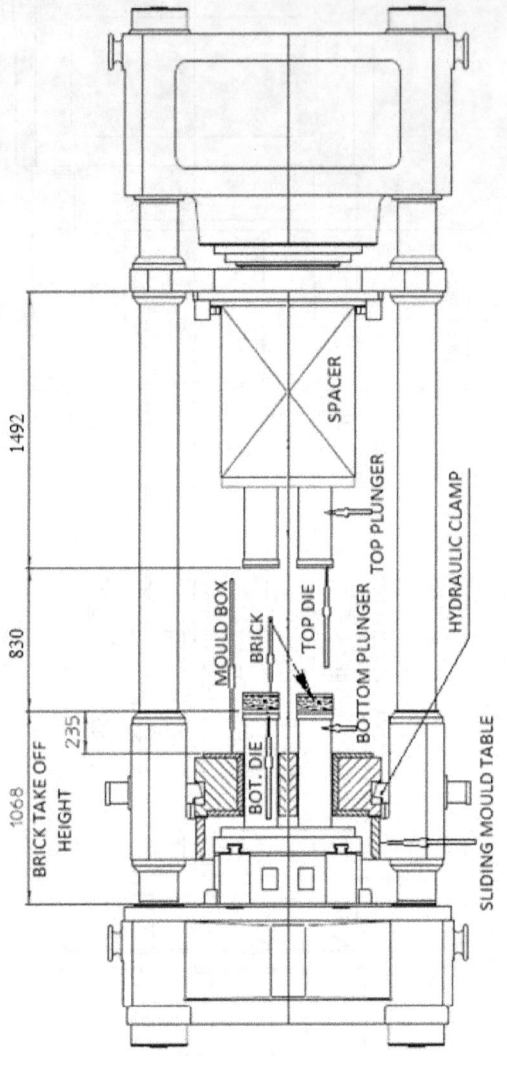

BRICK EJECTED

Chapter – 11

Die and plunger holding plate design

Design of dies:

Every brick has four sides and two faces. The features of bricks on four sides are maintained by providing liners on four sides of the cavity in mother mould. The features on top and bottom faces are maintained by providing dies.

Top Die: The die having negative features to give true shape or feature on the top pressing face of brick is called top die. The brick having no special feature on top surface will have plane die face.

Bottom Die: The die having negative features to give true shape or feature on bottom pressing face of brick is called bottom die. The brick having no any special feature on the bottom surface of brick will have plane die face.

Important considerations:

Top Die:

- The shape and size of the top die will match with an ultimate pressing zone of brick thickness in the mould cavity.
- The four sides will match with a feature on each individual side's liner. In the case of plain straight brick, sides of the die will have straight plain flat edges matching at the ultimate pressing zone.
- If open mould cavity has a key shape in the flat pressing system, the die will also have a matching key taper with the ultimate pressing zone.
- The side face of the die should have a matching negative profile with features in liners. It may be for the formation of tongue or groove or any profile on bricks.
- The surface of Die that will provide the top face of brick must match with the required feature in brick. It may be the plan, side taper in the case of Side Arch brick, End taper in the case of End Arch brick, side taper or end taper to form Key bricks in edge pressing system. It also may have any feature required on the top surface of the brick.
- It must have clearance on all sides to reduce pressure and friction on liners.
- It also must have tap holes of required size to fix with the plunger.

Bottom dies:

- The shape and size of the bottom die will match with the mould cavity at bottom pressing zone of brick thickness. It is at a distance of total filling depth.
- The four sides will match with a feature on each individual side liner. In the case of plain straight brick, sides of the die will have straight plain flat edge matching at pressing zone.

- If open mould cavity shape has a key shape in the flat pressing system, die will also have matching key shape with the ultimate pressing zone.
- The side face of the die may have a matching negative profile with features in liners. It may be for the formation of tongue or groove or any profile on bricks.
- The surface of Die that will provide the top face of brick must match with the required feature in brick. It may be the plane, side taper in the case of Side Arch brick, End taper in the case of End Arch brick, side taper or end taper to form Key bricks in edge pressing system. It also may have any feature requires on the bottom surface of the brick.
- It must have clearance on all sides to reduce pressure and friction on liners.
- It also must have tap holes of required size to fix with the plunger.

Dies for End Arch bricks:

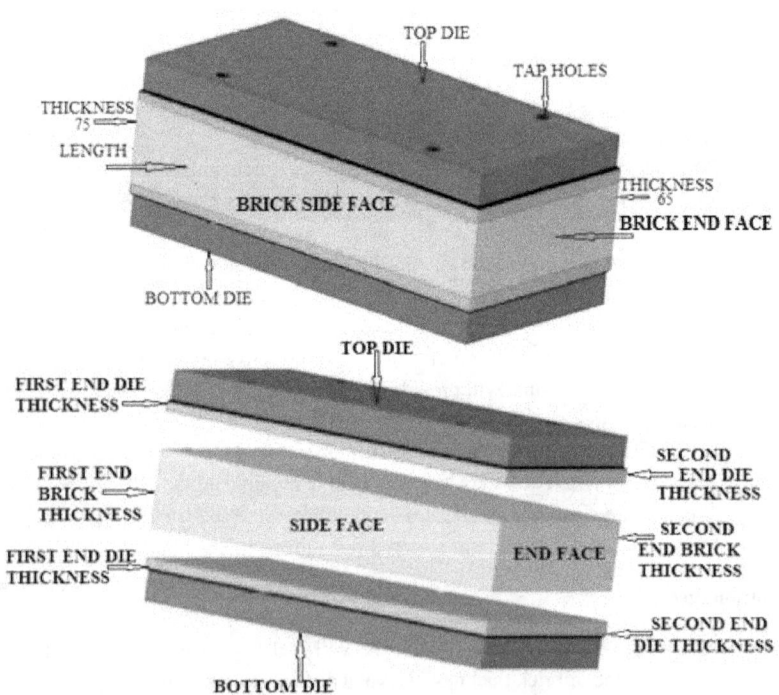

Dies for Side Arch brick:

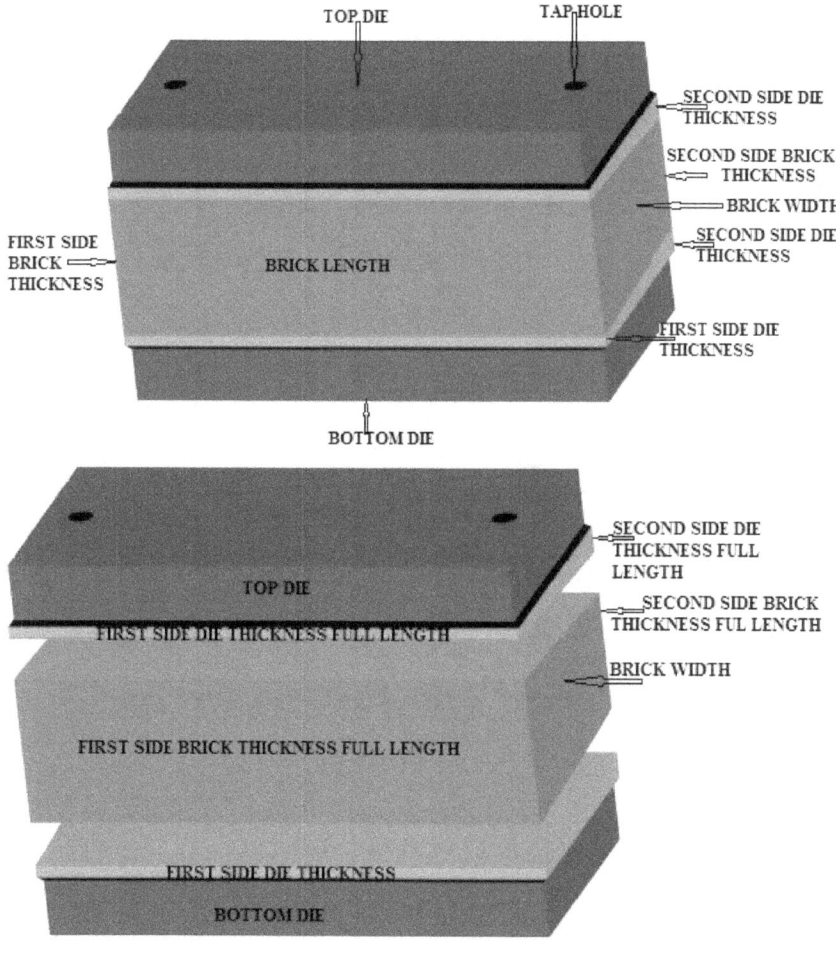

Dies for key bricks in flat pressing system:

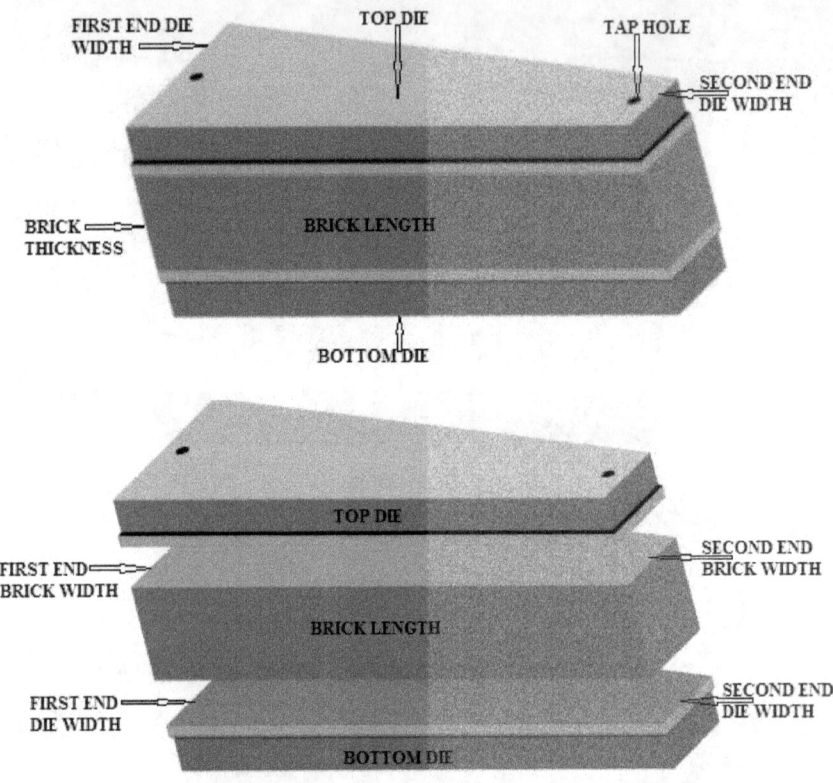

Dies for edge pressing Key brick:

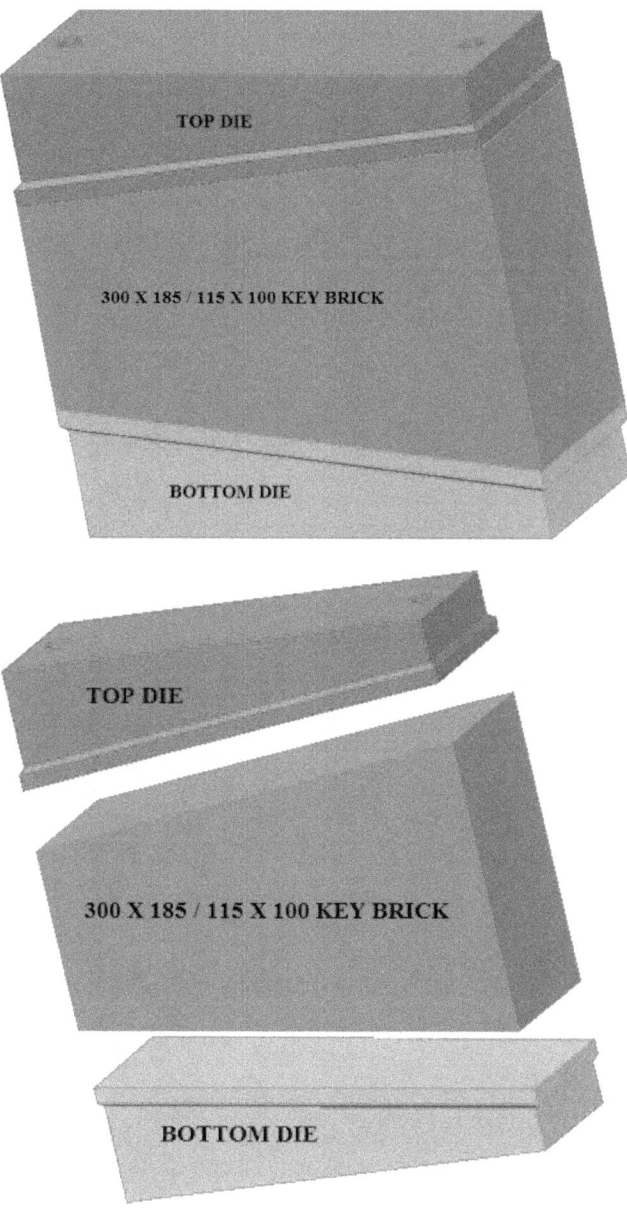

Die for Coke oven brick

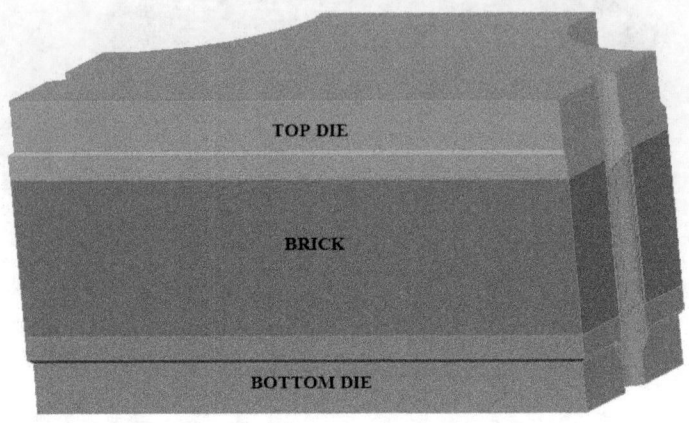

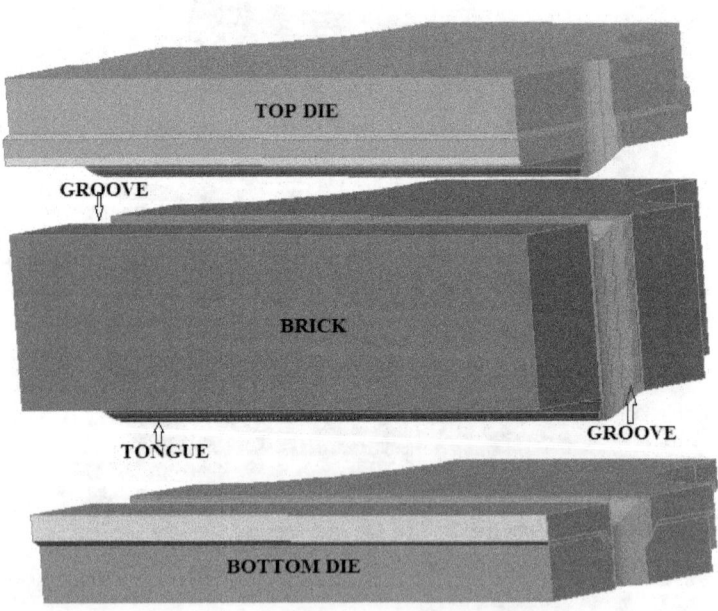

Design of plunger holding plates:

The plunger plates have very important role in brick manufacturing on a hydraulic press. It is classified as Top plunger holding plate and Bottom plunger holding plate.

The plate that connects top ram and top plunger with the top die are termed as Top Plunger holding-plate. Similarly, the plate that connects ram of bottom pressing cylinder or plunger holding-base with bottom plunger and bottom die is termed as bottom plunger holding-plate.

Organs of plunger holding plate

- Thick body of a steel plate having length and width of matching size with the plunger holding base at top and bottom Ram of the press.
- Drill holes at matching location and with matching size of Allen bolt to fix on holding-base.
- Tap holes at matching location with drill holes inside the pockets of the plungers. The size of tap hole must match with Allen bolt to be fitted

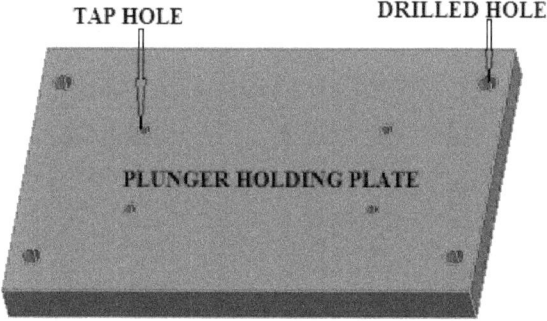

Plunger holding plate's surfaces should be machined, smooth, warpage-free and parallel. The centre lines of drill holes must match with tap holes of holding base. The tap holes on the plunger holding plate also must be at matching location and matching size of drill holes in plunger pockets.

If there is any other system of holding plunger plates in their respective position, the system of design must be studied from press manual and incorporated accordingly.

Chapter – 12

Brick shapes and sizes

Brick shapes for mould design:

The brick sizes written here may not be universal sizes. The size of brick depends on refractory lining design of individual company having their own standard.

Key bricks with variable length and key size:

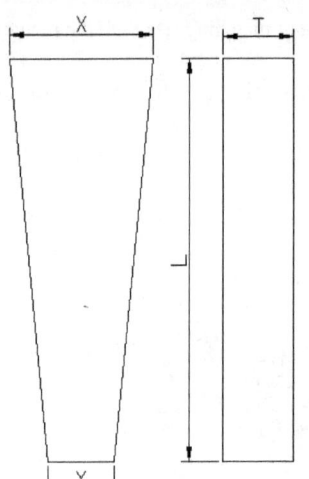

BRICK SHAPE	BRICK SIZE			
	L	X	Y	T
55/0	550	150	150	100
55/8	550	154	146	100
55/20	550	160	140	100
55/36	550	168	132	100
55/60	550	180	120	100
55/100	550	205	95	100

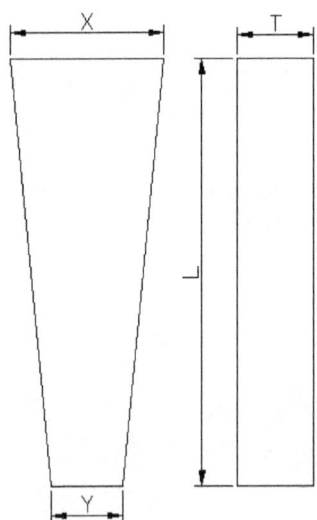

BRICK SHAPE	BRICK SIZE			
	L	X	Y	T
50/0	500	150	150	100
50/8	500	154	146	100
50/20	500	160	140	100
50/36	500	168	132	100
50/60	500	180	120	100
50/100	500	200	100	100

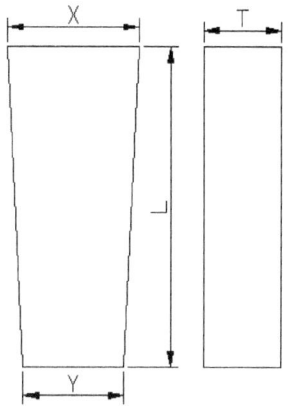

BRICK SHAPE	BRICK SIZE			
	L	X	Y	T
45/0	450	150	150	100
45/8	450	154	146	100
45/20	450	160	140	100
45/40	450	170	130	100
45/90	450	195	105	100

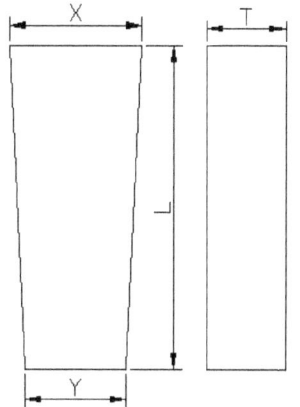

BRICK SHAPE	BRICK SIZE			
	L	X	Y	T
40/0	400	150	150	100
40/8	400	154	146	100
40/20	400	160	140	100
40/40	400	170	130	100
40/80	400	190	110	100

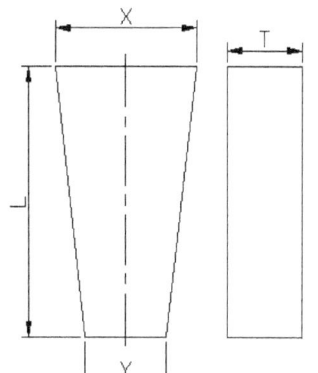

BRICK SHAPE	BRICK SIZE			
	L	X	Y	T
35/0	350	150	150	100
35/8	350	154	146	100
35/20	350	160	140	100
35/40	350	170	130	100
35/80	350	190	110	100

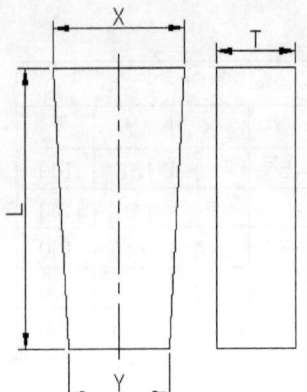

BRICK SHAPE	BRICK SIZE			
	L	X	Y	T
30/0	300	150	150	100
30/8	300	154	146	100
30/20	300	160	140	100
30/40	300	170	130	100
30/70	300	185	115	100

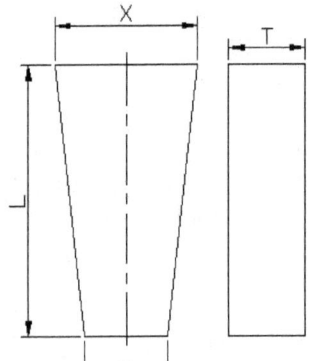

BRICK SHAPE	BRICK SIZE			
	L	X	Y	T
25/0	250	150	150	100
25/8	250	154	146	100
25/16	250	158	142	100
25/30	250	165	135	100
25/60	250	180	120	100

230X175/125X100
230X170/130X100
230X158/142X100
230X154/146X100

200X170/130X100
200X165/135X100
200X154/146X100
200X158/142X100

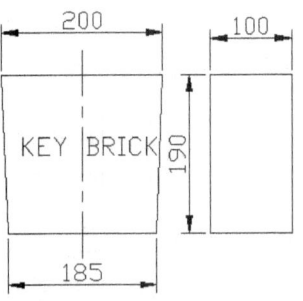

Side Arch and End Arch bricks:

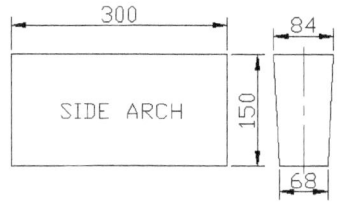

SIDE ARCH

300
84
150
68

300X150X101/51 SA
300X150X84/68 SA
300X150X88/64 SA
300X150X81/71 SA

300X225X75/65SA
345X114X79/73SA
375X124X69/59SA

250X155X105/95SA
250X155X110/90SA

250X250X105/95SA
230X100X75/68SA

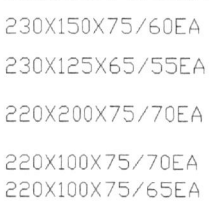

300X225X75/65EA

230X150X75/60EA

230X125X65/55EA

220X200X75/70EA

220X100X75/70EA
220X100X75/65EA

450X225X75/65EA

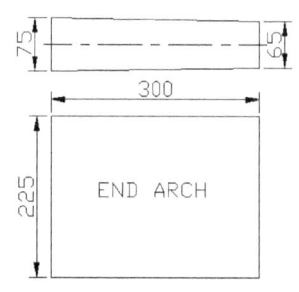

END ARCH

75
65
300
225

230 X 115 X 76

230 X 115 X 65

230 X 115 X 76 / 70 SA

232 X 117X 75

Coke oven brick

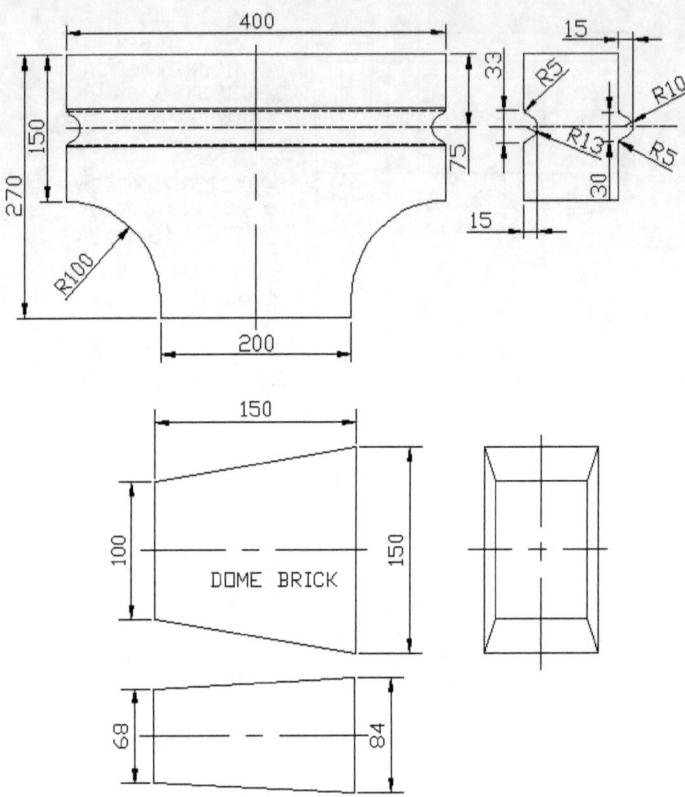

DOME BRICK

Circle brick:

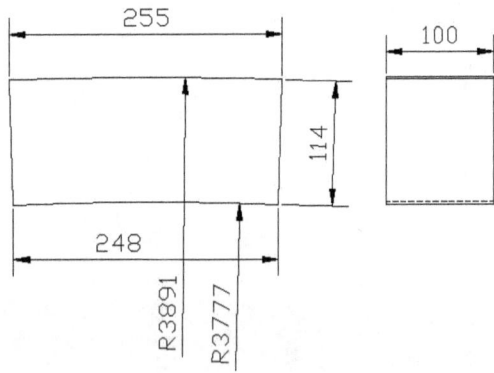

Chapter – 13

Practical guide on grouping

Grouping criteria:

- Fixed mother moulds for individual press considering pressing capacity.
- Group mother mould to accommodate shapes or items in the same cavity with variable packings.
- Form group for individual quality such as Silica, Basic, high Alumina, Firebrick
- The mother mould can be common if all above-said qualities are to be made in the common (same) press.
- Select items that can be made in same mould cavity: In this case, the top open surface area (length and width of the cavity) will be common. Brick pressing thickness will vary. It is applicable for side arch, end arch and key bricks.
- Consider individual group for Side Arch, End Arch and Key Bricks.
- Form separate group for flat pressing and edge pressing.
- Higher production demand of items must be considered. The customer should not suffer due to mother mould engaged in the production of another item.

Formation of group for moulding bricks, given in Chapter – 12

There are some standard sizes of bricks written in chapter-12 but these are not universal sizes. It depends on refractory lining design of individual company having their own standard.

Various sizes of key bricks, Side Arch, End Arch and shape bricks are there. Let us think there is only one press of 1000 tonnes capacity to produce all bricks. Consider the quality of brick is also same having no contraction or expansion in brick after firing. If there is any expansion or contraction provided by research and development laboratory, the layout for mould design must have these allowances on brick's basic sizes.

Consider Key brick sizes, Side Arch and End Arch brick sizes individually and think of mould design.

Selection of pressing Surface for grouping key bricks:

If we make a mould for flat pressing considering key shape surface for pressing, Pressing thickness will be the thickness of the brick. In this case, we have to design and make a mould for each key size having same length and thickness. See figure for flat pressing

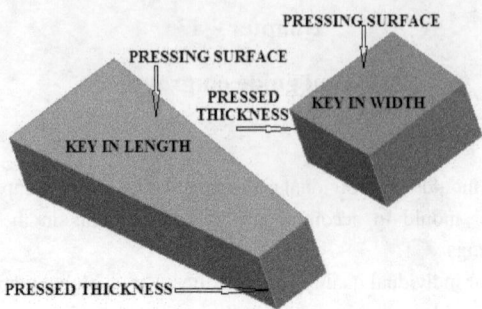

PRESSING SURFACE

PRESSING SURFACE

PRESSED THICKNESS

KEY IN WIDTH

KEY IN LENGTH

PRESSED THICKNESS

But if we think of making edge pressing considering length and thickness of brick as pressing surface, all key sizes in series can be made in single assembly by designing dies for individual key sizes. In edge pressing mould, height will be more than flat pressing but there will be saving against liners and assembly cost. The pressing surface will be as indicated in the figure.

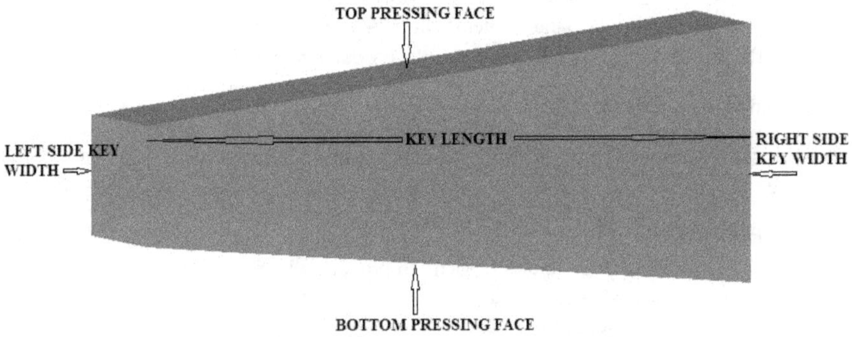

TOP PRESSING FACE

LEFT SIDE KEY WIDTH

KEY LENGTH

RIGHT SIDE KEY WIDTH

BOTTOM PRESSING FACE

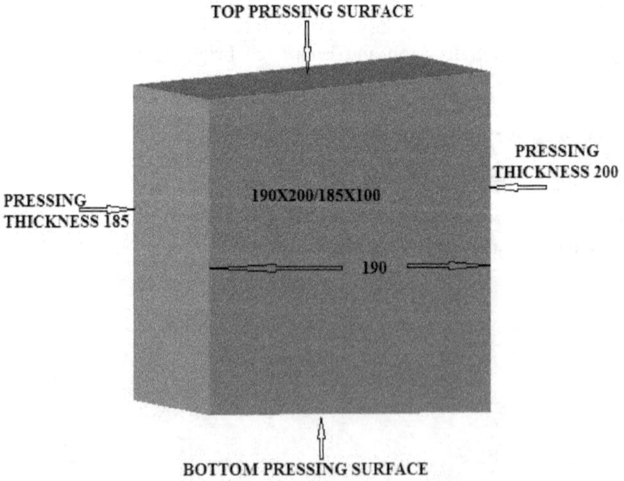

TOP PRESSING SURFACE

PRESSING THICKNESS 185

190X200/185X100

PRESSING THICKNESS 200

190

BOTTOM PRESSING SURFACE

The group formation is to be done considering variation in length and width. The possibility of assembly in mother mould cavity and suitability of press are also to be analysed. Let us consider required Specific pressure is 1 Tonne / cm^2, prepare the group considering the size and no of cavities that can be made.

Absolute Cavity Size (cm^2) = Press tonnage / required specific pressure in tonne

1000 tons / 1ton per cm^2= 1000 cm^2

No of cavity = Absolute cavity top surface area (Sq.cm) / Surface area of die face or open top surface area of one cavity in sq.cm.

Grouping for key bricks given in chapter – 12:

Series – 1

Serial No	Brick size (mm)	Mould open face (mm)	Filling depth (mm)	Mould height	No of cavity
1	550 X 150 X 100	550 X 100			
2	550 X 154/146 X 100	550 X 100			
3	550 X 160/140 X 100	550 X 100			
4	550 X 168/132 X 100	550 X 100			
5	550 X 180/120 X 100	550 X 100			
6	550 X 205/95 X 100	550 X 100	369	419	1

No of cavities = 1000 / (55*10) = 1.8, we can consider one cavity mould.

The calculated mould height is 419 but 400 mould height will be suitable to make bricks in above series. 205/95 is the maximum key size that is to be compressed between two dies. The single cavity common assembly with 400 mm mould height will produce all above-mentioned key sizes on 1000 tonnes hydraulic press. The mould cavity has to be laid lengthwise in mother mould cavity. Laying mould cavity width wise in mother mould will be at risk of a crack due to reduced wall thickness.

Series – 2

Serial No	Brick size (mm)	Mould open face (mm)	Filling depth (mm)	Mould height	No of cavity
1	500 X 150 X 100	500 X 100			
2	500 X 154/146 X 100	500 X 100			
3	500 X 160/140 X 100	500 X 100			
4	500 X 168/132 X 100	500 X 100			
5	500 X 180/120 X 100	500 X 100			
6	500 X 200/100 X 100	500 X 100	360	410	1

The calculated mould height is 410 but 400 mould height will be suitable to make bricks in above series. 200/100 mm is the maximum key size that is to be compressed between two dies. The single cavity common mould assembly in 400 mm mould height will produce all above-mentioned key sizes on 1000 tonnes hydraulic press. The mould cavity has to be laid lengthwise in mother mould cavity. Laying mould cavity width wise in mother mould will be at risk of a crack due to reduced wall thickness. The two cavities mould assembly will have full capacity utilisation of press. Required specific pressure may not be achieved.

Series – 3

Serial No	Brick size (mm)	Mould open face (mm)	Filling depth (mm)	Mould height	No of cavity
1	450 X 150 X 100	450 X 100			
2	450 X 154/146 X 100	450 X 100			
3	450 X 160/140 X 100	450 X 100			
4	450 X 170/130 X 100	450 X 100			
5	450 X 195/105 X 100	450 X 100	351	401	2

The calculated mould height is 401 but 400 mould height will be suitable to make bricks in above series. 195/105 is the maximum key size that is to be compressed between two dies. The two cavities common assembly with 400 mm mould height will produce all above-mentioned key sizes on 1000 tonnes hydraulic press. The calculated cavity size is 640*500 for 1000 Tonnes press with 1 tonne / square cm specific pressure. If the length of the brick is placed width wise the thickness of liner will be 25 mm and no packing is required.

This series can be in the group with series 1 and 2 or can go with series no 4 to 9.

Series – 4

Serial No	Brick size (mm)	Mould open face (mm)	Filling depth(mm)	Mould height	No of cavity
1	400 X 150 X 100	400 X 100			
2	400 X 154/146 X 100	400 X 100			
3	400 X 160/140 X 100	400 X 100			
4	400 X 170/130 X 100	400 X 100			
5	400 X 190/110 X 100	400 X 100	342	392	2

The calculated mould height is 392 but 400 mould height will be suitable to make bricks in above series. 190/110 is the maximum key size that is to be compressed between two dies. The two cavities common assembly with 400 mm mould height will produce all above-mentioned key sizes on 1000 tonnes hydraulic press. The mould cavities can be laid widthwise in mother mould cavity.

Series – 5

Serial No	Brick size (mm)	Mould open face (mm)	Filling depth(mm)	Mould height	No of cavity
1	350 X 150 X 100	350 X 100			
2	350 X 154/146 X 100	350 X 100			
3	350 X 160/140 X 100	350 X 100			
4	350 X 170/130 X 100	350 X 100			
5	350 X 190/110 X 100	350 X 100	342	392	2

The calculated mould height is 392 but 400 mould height will be suitable to make bricks in above series. 190/110 is the maximum key size that is to be compressed between two dies. The two cavities common assembly with 400 mm mould height will produce all above-mentioned key sizes on 1000 tonnes hydraulic press. The mould cavities can be laid widthwise in mother mould cavity.

Series – 6

Serial No	Brick size (mm)	Mould open face (mm)	Filling depth (mm)	Mould height	No of cavity
1	300 X 150 X 100	300 X 100			
2	300 X 154/146 X 100	300 X 100			
3	300 X 160/140 X 100	300 X 100			
4	300 X 170/130 X 100	300 X 100			
5	300 X 185/115 X 100	300 X 100	333	383	3

The calculated mould height is 383 but 400 mould height will be suitable to make bricks in above series. 185/115 is the maximum key size that is to be compressed between two dies. The three cavities common assembly with 400 mm mould height will produce all above-mentioned key sizes on 1000 tonnes hydraulic press. The mould cavities to be laid widthwise in mother mould cavity.

Series – 7

Serial No	Brick size (mm)	Mould open face (mm)	Filling depth (mm)	Mould height	No of cavity
1	250 X 150 X 100	250 X 100			
2	250 X 154/146 X 100	250 X 100			
3	250 X 158142 X 100	250 X 100			
4	250 X 160/140 X 100	250 X 100			
5	250 X 170/130 X 100	250 X 100			
6	250 X 185/115 X 100	250 X 100	333	383	3

The calculated mould height is 383 but 400 mould height will be suitable to make bricks in above series 185/115 is the maximum key size that is to be compressed between two dies. The three cavities common assembly with 400 mm mould height will produce all above-mentioned key sizes on 1000 tonnes hydraulic press. The mould cavities to be laid widthwise in mother mould cavity.

Series – 8

Serial No	Brick size (mm)	Mould open face (mm)	Filling depth (mm)	Mould height	No of cavity
1	230 X 150 X 100	230 X 100			
2	230 X 154/146 X 100	230 X 100			
3	230 X 158142 X 100	230 X 100			
4	230 X 160/140 X 100	230 X 100			
5	230 X 170/130 X 100	230 X 100	306	356	4

The calculated mould height is 356 but 400 mould height can be used to make bricks in above series. 170/130 is the maximum key size that is to be compressed between two dies. The 4 cavities moulds are not possible in the mother mould cavity. The three cavities common assembly with 400 mm mould height will produce all above-mentioned key sizes on 1000 tonnes hydraulic press. The mould cavities to be laid widthwise in mother mould cavity.

Series – 9

Serial No	Brick size (mm)	Mould open face (mm)	Filling depth (mm)	Mould height	No of cavity
1	200 X 150 X 100	200 X 100			
2	200 X 154/146 X 100	200 X 100			
3	200 X 158/142 X 100	200 X 100			
4	200 X 160/140 X 100	200 X 100			
5	200 X 170/130 X 100	200 X 100	306	356	4

The calculated mould height is 356 but 400 mould height can be used to make bricks in above series. 170/130 is the maximum key size that is to be compressed between two dies. The three cavities common assembly with 400 mm mould height will produce all above-mentioned key sizes on 1000 tonnes hydraulic press. The mould cavities to be laid widthwise in mother mould cavity.

The absolute cavity size in mother mould will be 640 X 500 mm with 1 tonne / cm^2 specific pressure on 1000 tonnes pressing capacity of the press having 1050 length and 910 width of mother mould. The 4 cavity mould could not be made as cavity length will be more than the absolute cavity length. The mother mould may crack due to reduced wall thickness.

The mould cavity for making all bricks from series 1 to 3 have to be laid lengthwise along the length of mother mould for making on this press. Laying multi-cavity mould along width wise will be a better option on the higher capacity press.

The items from series 4 to 9 can be made by laying multi-cavity mould width wise on this press. The mould height from series 4 to 7 can be 400

Calculated cavity size in mother mould:

The series written in the table represents all sizes of bricks given under in series no earlier.

Series No	Die face	Mould height	No of cavity	Liner thickness	Packing thickness	Middle packing	Mould cavity Length	width
4	400 X 100	392	2	29	0	14	330	458
5	350 X 100	392	2	29	0	14	330	408
6	300 X 100	383	3	29	0	14	502	358
7	250 X 100	383	3	29	0	14	502	308
8	230 X 100	356	4	29	0	14	674	288
9	200 X 100	356	4	29	0	14	674	258

Cavity length = (Brick width at top*No of Cavities) + {(middle packing thickness* (No of cavities – 1)} + (liner thickness at top* No of cavities*2) + (End packing thickness*2)

Cavity width = Brick length + (liner thickness at top*2) + (side packing thickness*2)

The cavity size written in above table has been calculated on the basis of above equation and has been corrected as in the table given below:

Corrected Cavity size, mould height and packing thickness:

Series No	Die face	Mould height	No of cavity	Liner thickness	Packing thickness E/P	S/P	Middle packing	Mould cavity Length	width
4	400 X 100	400	2	29	110	21	14	550	500
5	350 X 100	400	2	29	110	46	14	550	500
6	300 X 100	400	3	29	30	21	14	562	400
7	250 X 100	400	3	29	30	46	14	562	400
8	230 X 100	400	3	29	30	56	14	562	400
9	200 X 100	400	3	29	30	71	14	562	400

End Packing thickness = [Length of cavity - {(Brick width at the top of mould cavity * No of cavities) + {(Thickness of middle packing * (No of cavities - 1)} + (Thickness of liner at the top side of cavity * No of cavities * 2)}] / 2

Side Packing thickness = Width of cavity in mother mould – (Brick length at the top of mould cavity + Liner thickness at top side * 2) / 2

In series 4 and 5 the mould cavity has been fixed as mentioned in above table and packing thickness calculated on the basis of above equation. There are two set of mother moulds having a difference in cavity size but mould height will be 400 mm for all.

The four cavity mould with 1 tonne /cm^2 specific pressure for series 8 and 9 cannot be made on this press in the mother mould of size 1050 X 910 as cavity length will be large enough, making the weak cross section.

Note: The group is always made as per manufacturing orders and convenient of the company. There is no any formula or hard and fast rule for grouping. You can follow the guidelines provided earlier for convenient.

Grouping for Side Arch bricks:

The flat pressing mould assembly in same mother mould will be economical if bricks of variable side arch having common length and width can be made in the common mould assembly. Make a table of brick sizes with possibilities of cavity formation.

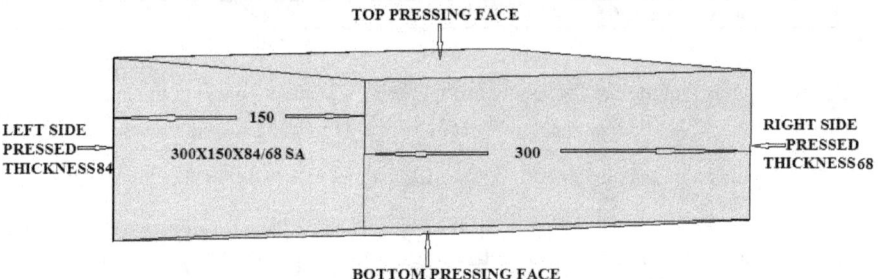

Serial No	Brick size (mm)	Mould open face (mm)	Filling depth (mm)	Mould height	No of cavity	Mother mould cavity
1	300X150X101/51SA	300 X 150	181.8	231.5	2	500 X400
2	300X150X84/68 SA	300 X 150	151.2	201.2	2	
3	300X150X88/64 SA	300 X 150	158.4	208.4	2	
4	300X150X81/71SA	300 X 150	145.8	195.8	2	
5	300X150X101/51SA	300 X 150	181.8	231.8	2	
6	300X225X75/65 SA	300 X 225	135	185	1	500 X 400
7	345X114X79/73 SA	345 X 114	131.4	181.4	2	500 X 450
8	375X124X69/59 SA	375 X 124	124.2	174.2	2	500 X 450
9	250X155X105/95SA	250 X 155	189	239	2	500 X 400
10	250X155X110/90SA	250 X 155	198	248	2	
11	250X250X105/95SA	250 X 250	189	239	1	500 X 400
12	230X100X75/68 SA	230 X 100	135	185	4	674 X 400

There are 12 individual sizes of Side arch bricks in above table. In the total group of 12 items lowest required mould height is 174.2 and highest 248 mm. The common mother mould height for all items in the above group can have 250 mm. All items from serial no 1 to 5 can be made in single assembly. Similarly, item 9 and 10 can be made in common assembly. The required cavity length for brick size serial no 12 in above table for 4 cavities will be 674+. The available cross section of wall at both ends will be less than a requirement. We can make

3 cavities mould in mother mould with 600 X 400 cavity size used for making end arch bricks.

Corrected Cavity size, mould height and packing thickness in above table to have common mother mould:

Serial No	Die face	Mould height	No of cavity	Liner thickness	Packing thickness		Middle packing	Mould cavity	
					E/P	S/P		Length	width
1	300 X 150	250	2	29	35	21	14	500	400
2	300 X 150	250	2	29	35	26	14	500	400
3	300 X 150	250	2	29	35	26	14	500	400
4	300 X 150	250	2	29	35	26	14	500	400
5	300 X 150	250	2	29	35	26	14	500	400
6	300 X 225	250	1	29	71	58.5	0	500	400
7	345 X 114	250	2	29	71	23.5	14	500	450
8	375 X 124	250	2	29	61	8.5	14	500	450
9	250 X 155	250	2	29	30	46	14	500	400
10	250 X 155	250	2	29	30	46	14	500	400
11	250 X 250	250	1	29	96	46	0	500	400
12	230 X 100	250	3	29	0	51	13	500	400

Grouping for End Arch Bricks

The flat pressing mould assembly in same mother mould will be economical if bricks of variable end arch having common length and width can be made in the common mould assembly. Make table of brick sizes with possibilities of cavity formation

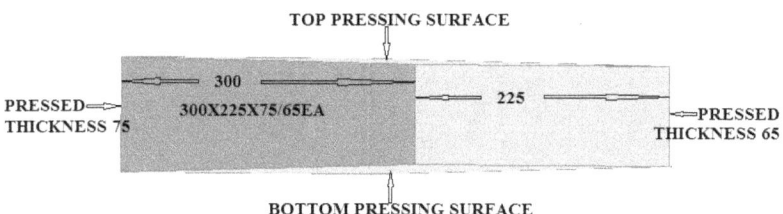

Serial no	Brick size (mm)	Mould open face (mm)	Filling depth (mm)	Mould height	No of cavity	Mother mould cavity
1	450X225X75/65 EA	450 X 225	135	185	1	600 X 400
2	300X225X75/65 EA	300 X 225	135	185	1	500 X 400
3	230X150X75/60 EA	230X 150	135	185	2	500 X 400
4	230X125X65/55 EA	230 X 125	117	167	3	600 X 400
5	220X200X75/70 EA	220 X 200	135	185	2	600 X 400
6	220X100X75/70 EA	220X 100	135	185	4	600 X 400
7	220X100X75/65 EA	220X 100	135	185	4	600 X 400

There are 7 brick sizes in above table. The brick sizes of serial no 6 and 7 can be made in a common assembly with a separate set of dies. The Cavities in mother mould for the item in serial No 1, 4, 5, 6 and 7 are of common size. Similarly, items of serial No 2 and 3 have common cavity size in mother mould. See the corrected size of mould height packing plate thickness and mother mould cavity in the table given below:

Corrected Cavity size, mould height and packing thickness:

Series No	Die face	Mould height	No of cavity	Liner thickness	Packing thickness		Middle packing	Mould cavity	
					E/P	S/P		Length	width
1	450 X 225	250	1	29	46	58.5	0	600	400
2	300 X 225	250	1	29	71	58.5	0	500	400
3	230 X 150	250	2	29	35	56	14	500	400
4	230 X 125	250	3	29	11.5	56	14	600	400
5	220 X 200	250	2	29	35	61	14	600	400
6	220 X 100	250	3	29	49	61	14	600	400
7	220 X 100	250	3	29	49	61	14	600	400

The moulding of all above items can be done in two mother moulds with a set of liners. The mould heights for all items have 250 mm height. The cavity size for item 2 and 3 is common; similarly cavity size for item 1, 4, 5, 6 and 7 is common.

Chapter – 14

Edge pressing moulds for making key bricks

Design mother mould and all accessories required for making key bricks on 1000 Tonnes press having daylight and working parameter as given earlier. The size of bricks and permissible tolerance and taper is given here.

(1) 300 X 150 X 100 (2) 300 X 154 / 146 X 100 (3) 300 X 160 / 140 X 100

(4) 300 X 170 /130 X 100 (5) 300 X 185 X / 115 X 100

Tolerance $\pm {}^{1.0}_{2.0}$

The taper on ejected surface (surface coming in contact to liners) = 1.0 Maximum

The above-given sizes of bricks are from chapter – 13 groups of key bricks series – 6. The size of required mother mould is already given in the table. The mother mould has been designed for making bricks on 100 tonnes hydraulic press. The location of tap holes and size has been designed to match with top and bottom lock plates.

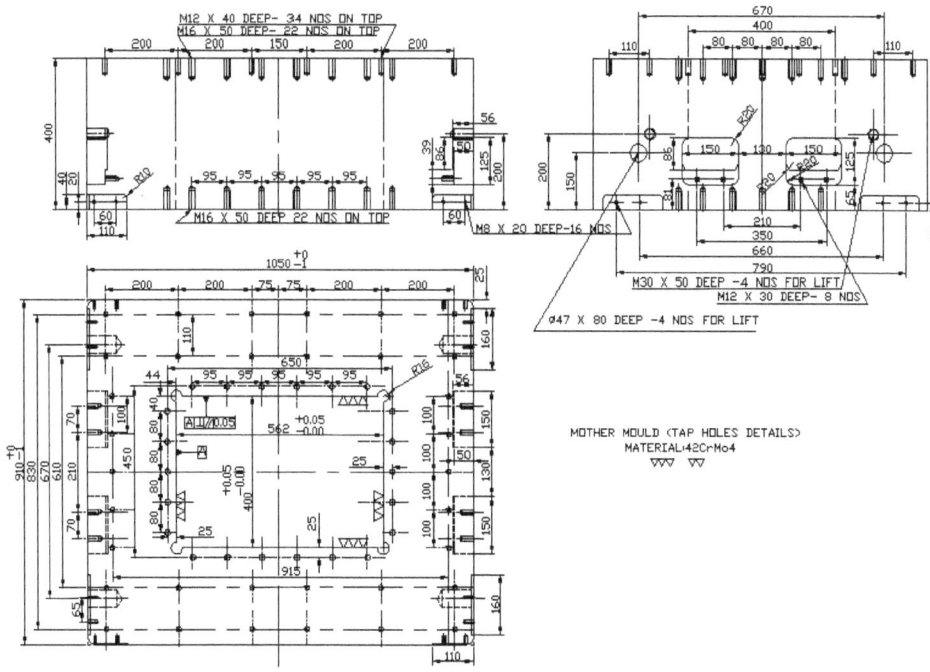

MOTHER MOULD (TAP HOLES DETAILS)
MATERIAL:42Cr-Mo4

The mould height and cavity size have been made for adequate filling depth and to accommodate liners and packings for 3 cavity moulds as calculated in the table of series – 6.

The design plans for 3 items are given to explain the process of design.

Design plan for 300 X 150 X 100:

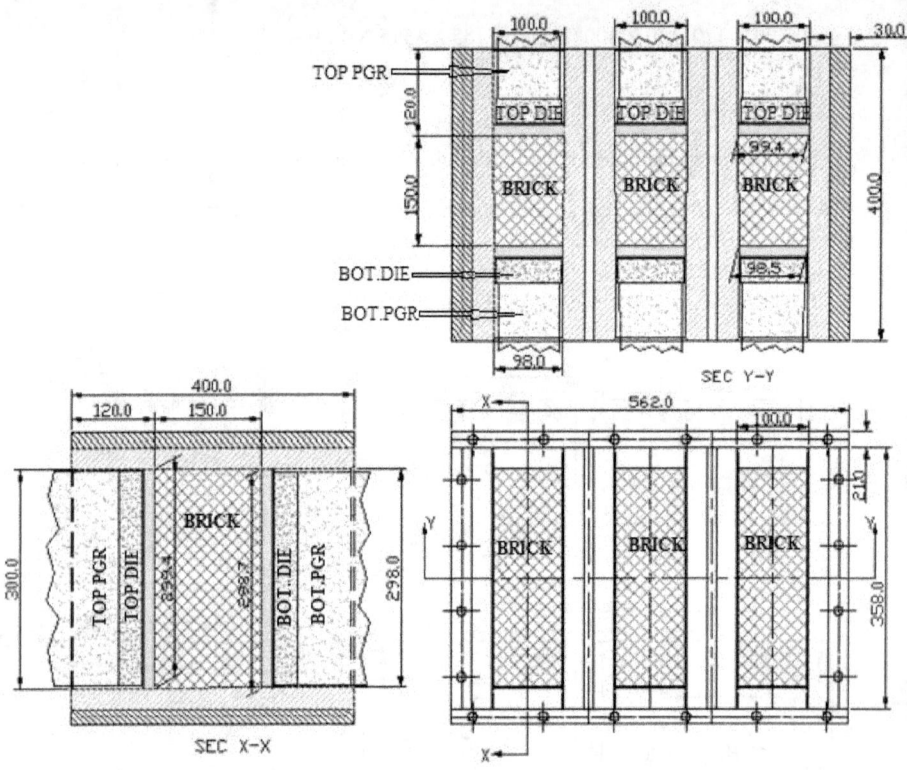

Design plan for 300 X 154 / 146 X 100:

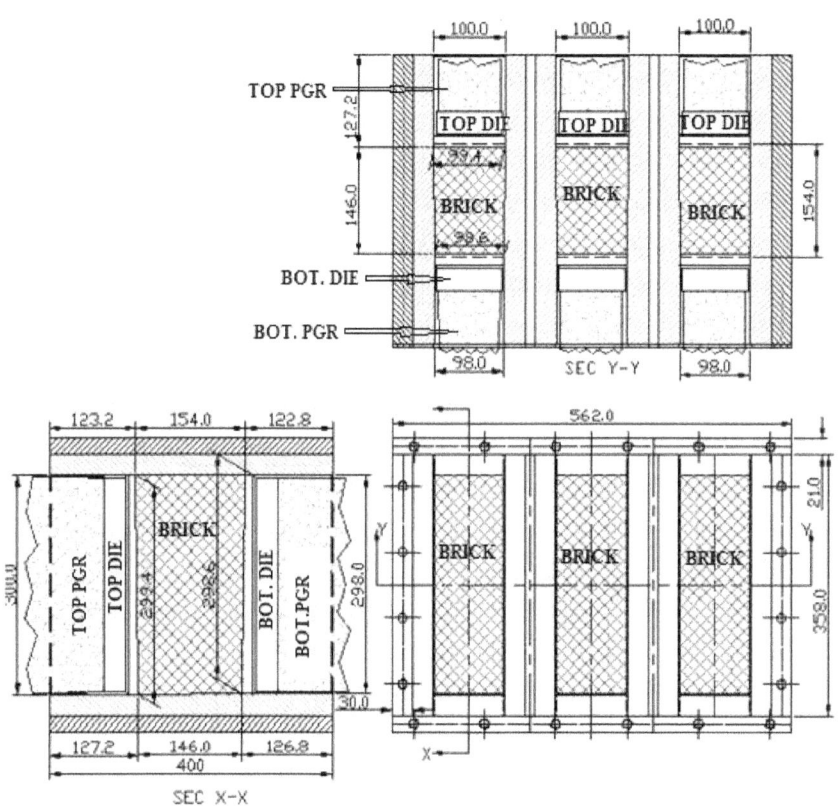

Design plan for 300 X 185 / 115 X 100:

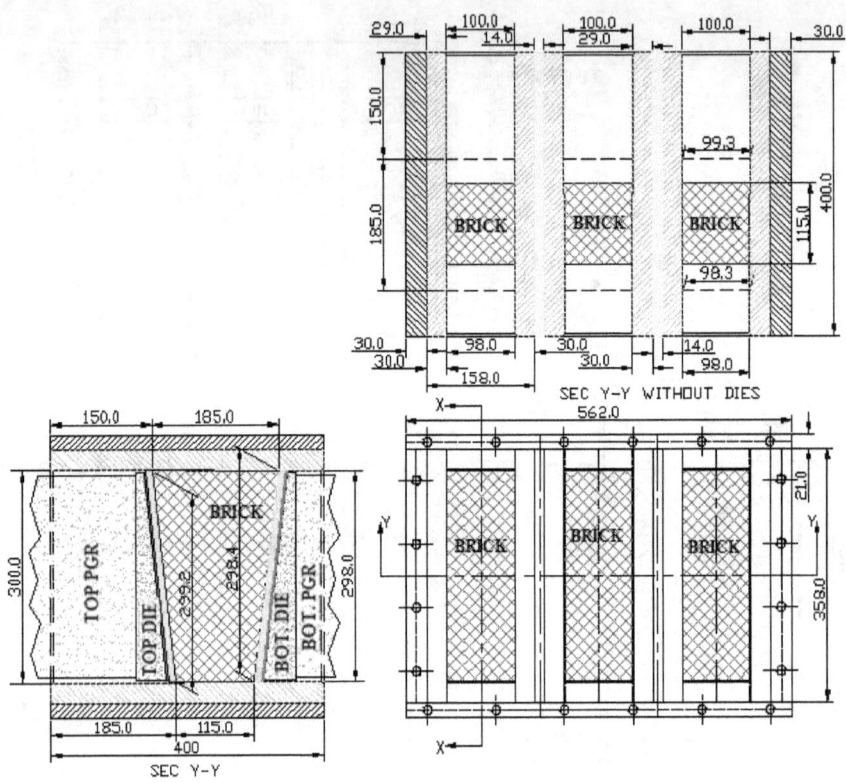

Design of side liner, end liners, side packing, end packing, dies and plungers:

The drawings of side liner, end liners, packing plates, dies and plungers have been drawn as per designed plan layout. The liners, plungers, and packing are common for all 5 items. The top and bottom dies have been designed to match with the key size of bricks. The individual drawings for all items are given here to impart training to learners in mould design.

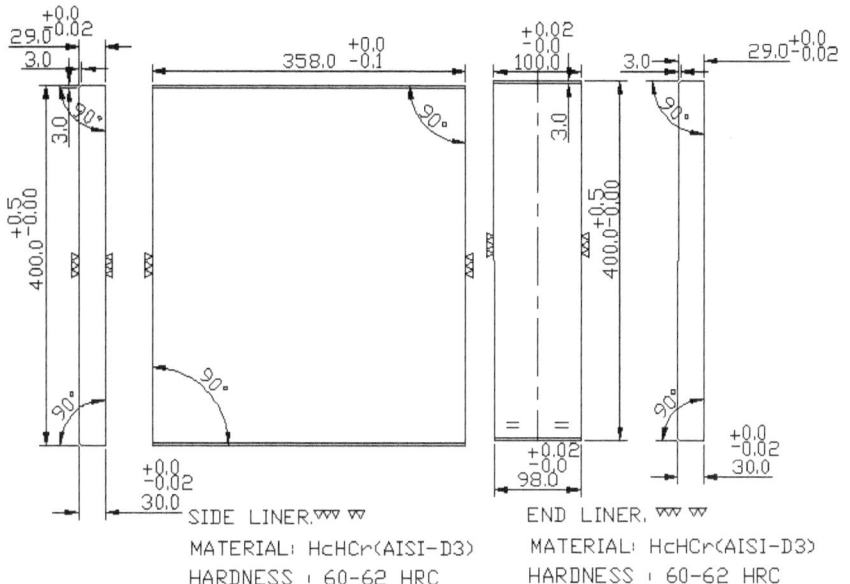

SIDE LINER. ▽▽▽ ▽
MATERIAL: HcHCr(AISI-D3)
HARDNESS : 60-62 HRC
QUANTITY: 6 NOS (3 SETS)

END LINER, ▽▽▽ ▽
MATERIAL: HcHCr(AISI-D3)
HARDNESS : 60-62 HRC
QUANTITY: 6 NOS (3 SETS)

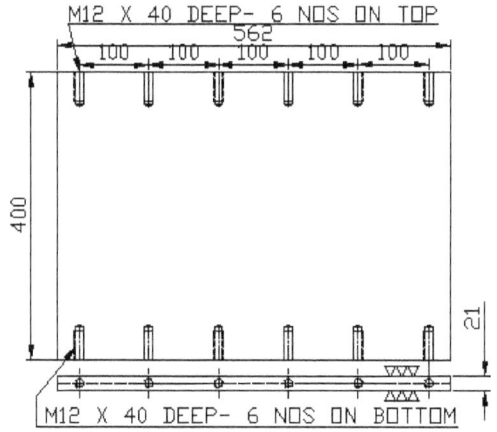

M12 X 40 DEEP- 6 NOS ON TOP

M12 X 40 DEEP- 6 NOS ON BOTTOM

SIDE PACKING ▽▽▽ ▽▽
MATERIAL: EN -19
HARDNESS: 40 - 42 HRC
QUANTITY: 2 NOS

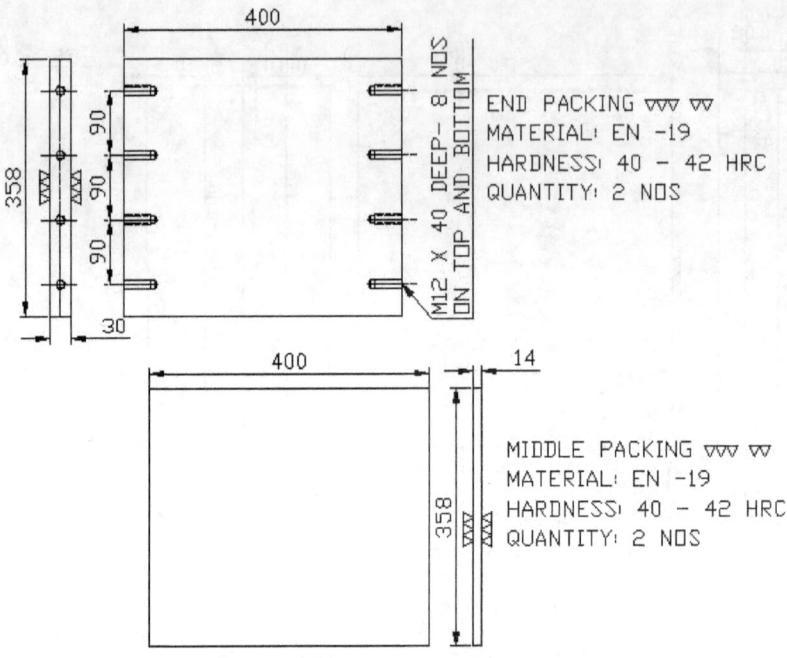

END PACKING ▽▽▽ ▽▽
MATERIAL: EN -19
HARDNESS: 40 - 42 HRC
QUANTITY: 2 NOS

MIDDLE PACKING ▽▽▽ ▽▽
MATERIAL: EN -19
HARDNESS: 40 - 42 HRC
QUANTITY: 2 NOS

Dies for 300 X 150 X100 brick:

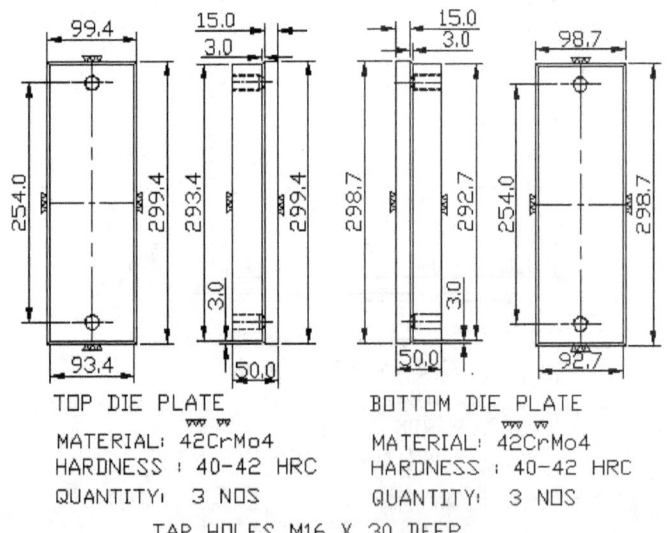

TOP DIE PLATE ▽▽ ▽▽
MATERIAL: 42CrMo4
HARDNESS : 40-42 HRC
QUANTITY: 3 NOS

BOTTOM DIE PLATE ▽▽ ▽▽
MATERIAL: 42CrMo4
HARDNESS : 40-42 HRC
QUANTITY: 3 NOS

TAP HOLES M16 X 30 DEEP

Dies for 300 X 154/ 146 X 100 Key brick:

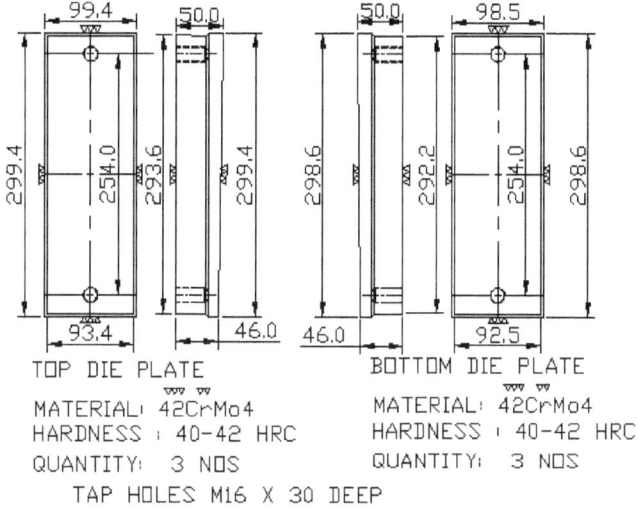

TOP DIE PLATE
MATERIAL: 42CrMo4
HARDNESS : 40-42 HRC
QUANTITY: 3 NOS

BOTTOM DIE PLATE
MATERIAL: 42CrMo4
HARDNESS : 40-42 HRC
QUANTITY: 3 NOS

TAP HOLES M16 X 30 DEEP

Dies for 300 X 160 / 140 X 100 Key brick:

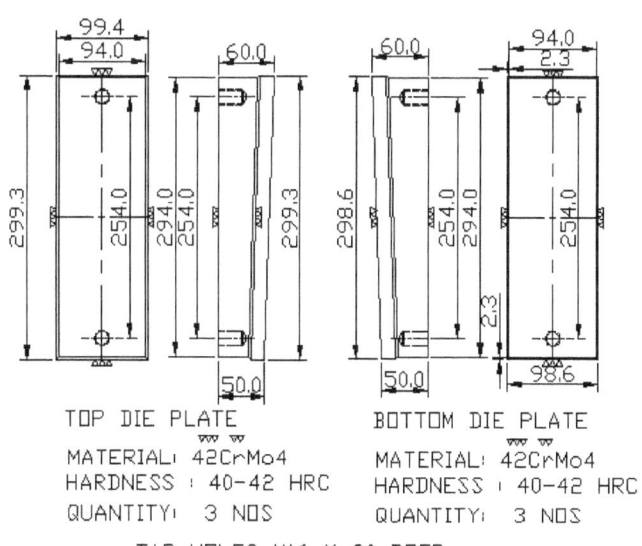

TOP DIE PLATE
MATERIAL: 42CrMo4
HARDNESS : 40-42 HRC
QUANTITY: 3 NOS

BOTTOM DIE PLATE
MATERIAL: 42CrMo4
HARDNESS : 40-42 HRC
QUANTITY: 3 NOS

TAP HOLES M16 X 30 DEEP

Dies for 300 X 170 / 130 X 100 Key brick:

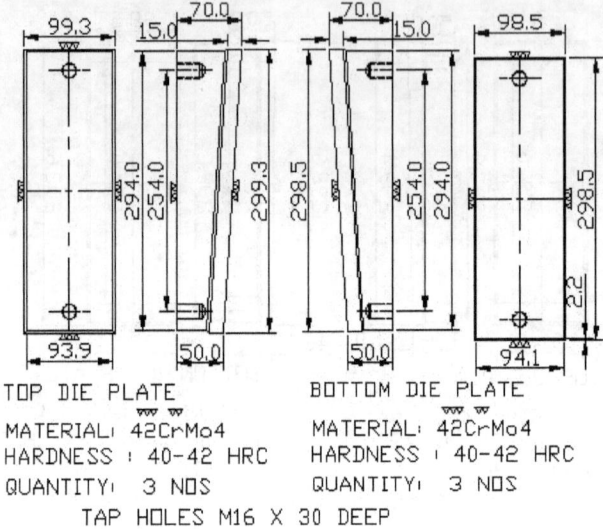

TOP DIE PLATE
MATERIAL: 42CrMo4
HARDNESS : 40-42 HRC
QUANTITY: 3 NOS

BOTTOM DIE PLATE
MATERIAL: 42CrMo4
HARDNESS : 40-42 HRC
QUANTITY: 3 NOS

TAP HOLES M16 X 30 DEEP

Die for 300 X 185 / 115 X 100 Key brick:

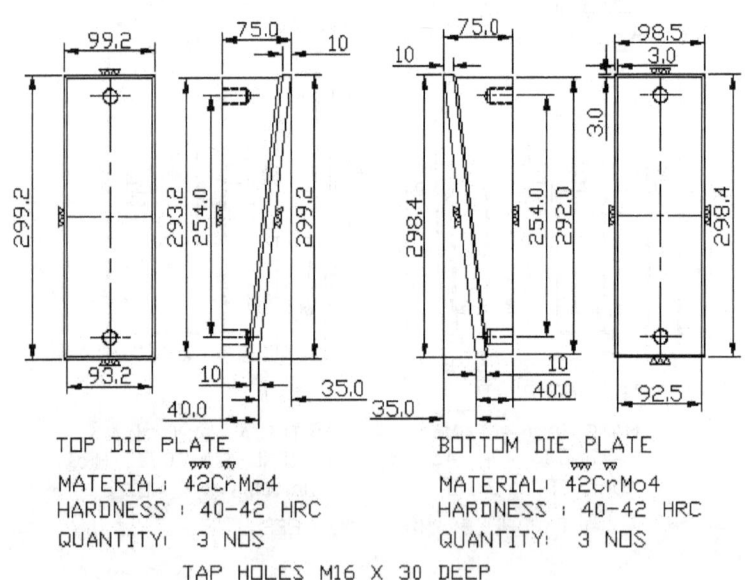

TOP DIE PLATE
MATERIAL: 42CrMo4
HARDNESS : 40-42 HRC
QUANTITY: 3 NOS

BOTTOM DIE PLATE
MATERIAL: 42CrMo4
HARDNESS : 40-42 HRC
QUANTITY: 3 NOS

TAP HOLES M16 X 30 DEEP

Note: The material and hardness of all accessories can be changed to prolong the life of mould accessories.

Top lock plate, Bottom lock plate and cover plates:

The counter bore holes in lock plates are to connect with tap hole on top and bottom face of mother mould. The counter bores represented by red lines will attach packing plates with it and others with mother mould. The cover plates will be fitted on the top surface of mother mould for smooth movement of the charger. The location of counter bores in lock plates will vary according to the location of tap holes in mother mould and packing plates.

The cavities in lock plates should be bigger than open cavity surface at top and bottom face of the assembly. It is generally taken as 2-3 mm on each side approximately matching with chamfer lines.

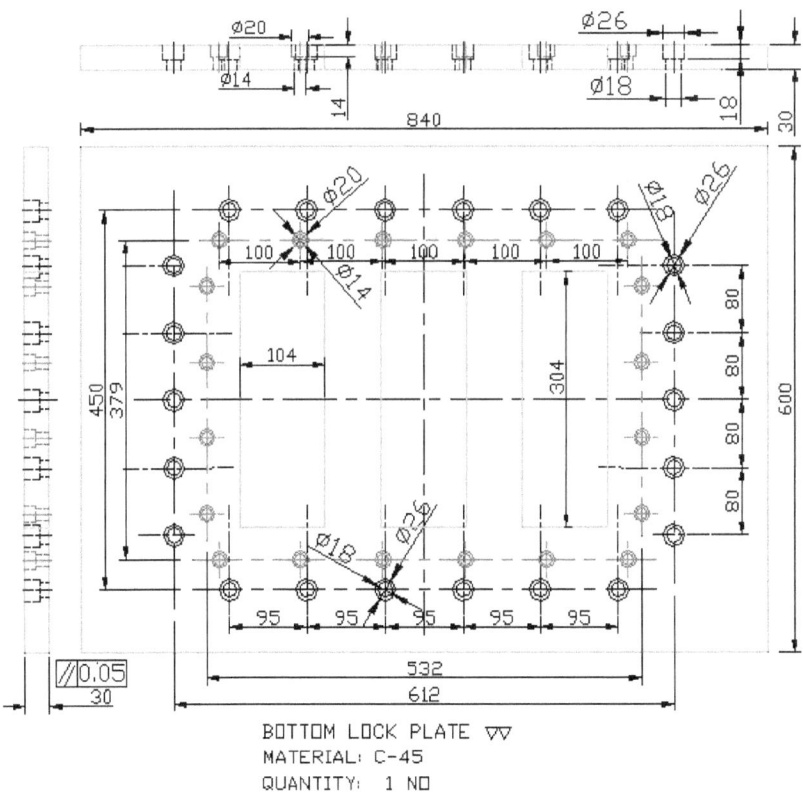

BOTTOM LOCK PLATE ▽▽
MATERIAL: C-45
QUANTITY: 1 NO

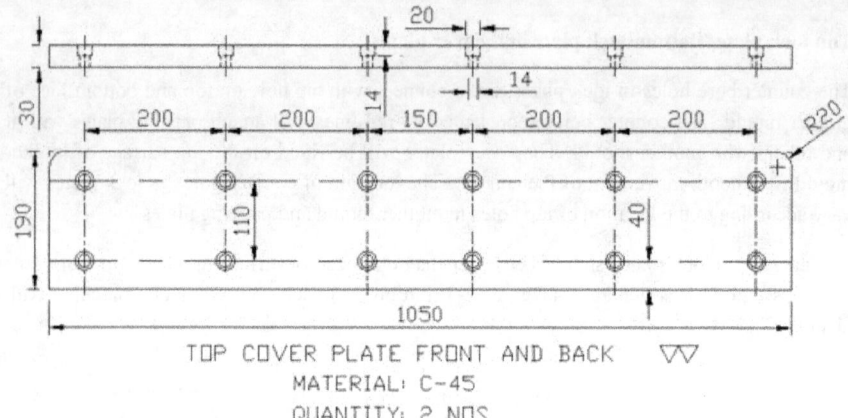

TOP COVER PLATE FRONT AND BACK ▽▽
MATERIAL: C-45
QUANTITY: 2 NOS

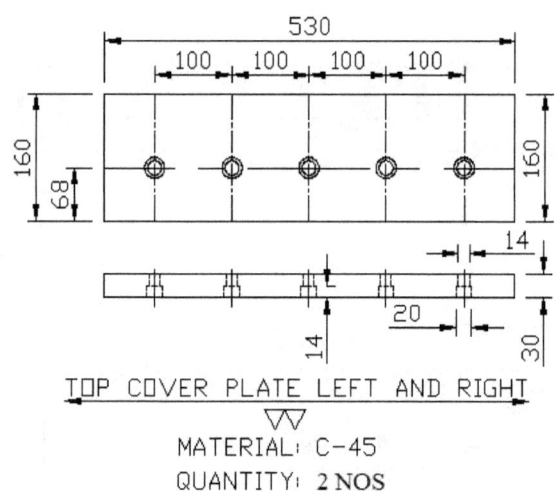

TOP COVER PLATE LEFT AND RIGHT
▽▽
MATERIAL: C-45
QUANTITY: 2 NOS

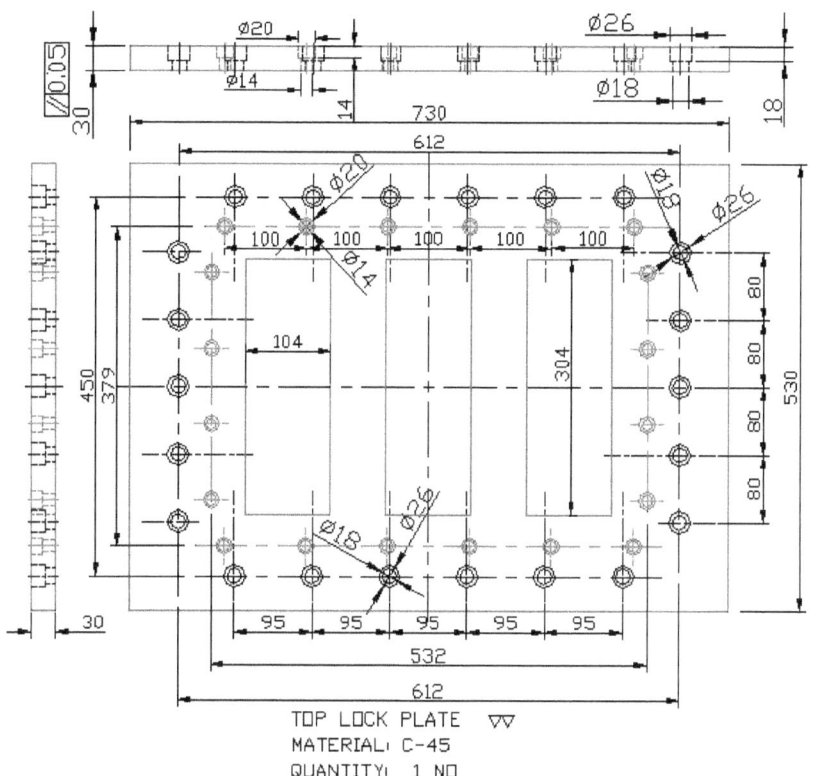

TOP LOCK PLATE ▽▽
MATERIAL: C-45
QUANTITY: 1 NO

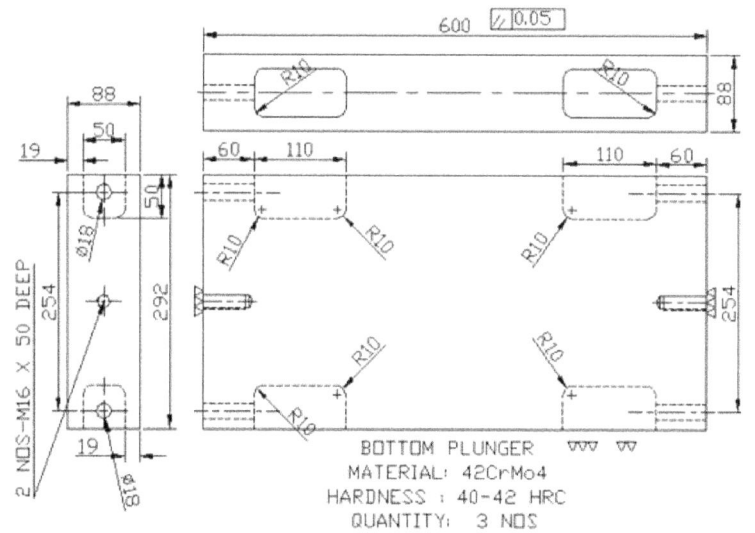

BOTTOM PLUNGER ▽▽▽ ▽▽
MATERIAL: 42CrMo4
HARDNESS : 40-42 HRC
QUANTITY: 3 NOS

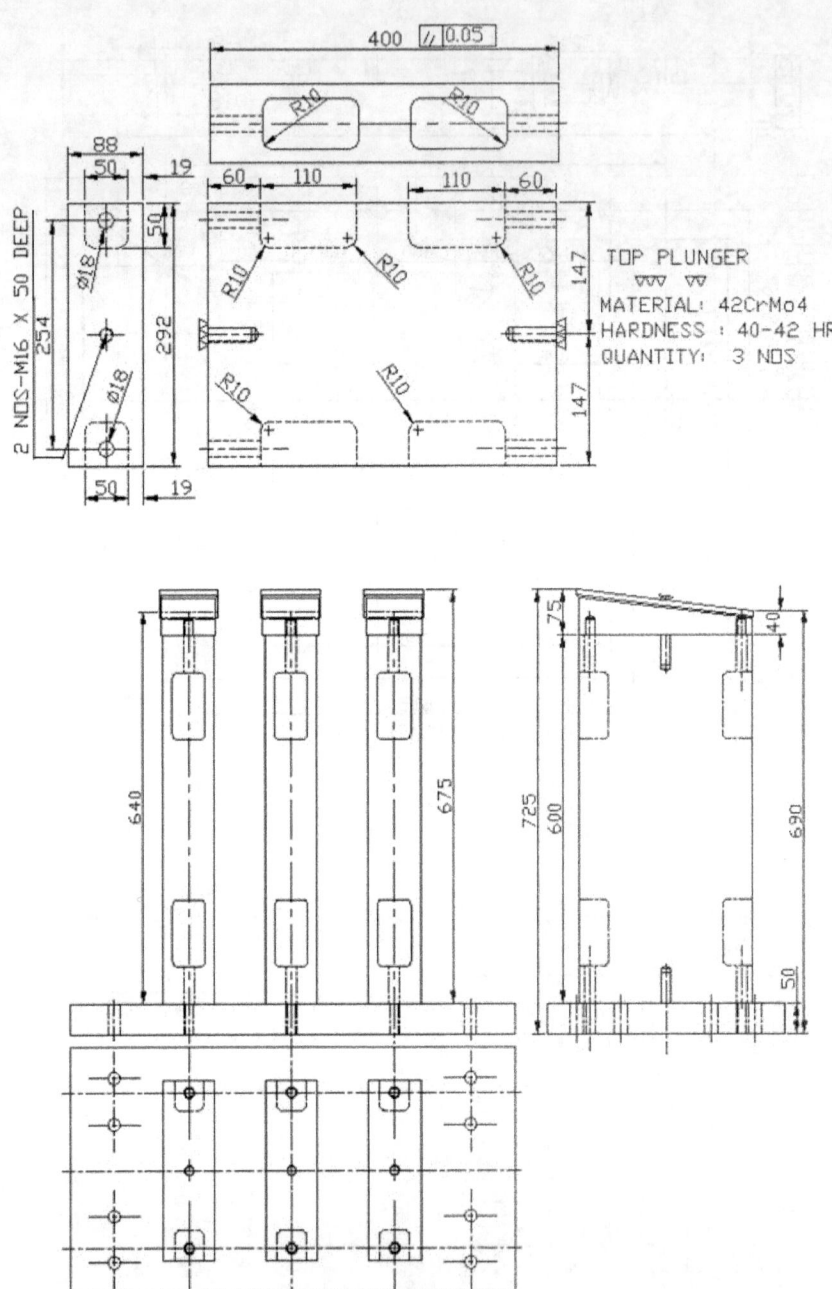

400   // 0.05

R10   R10

88
50   19
254
292
50
Ø18
Ø18
2 NOS-M16 X 50 DEEP

60   110   110   60
147
R10   R10   R10
R10   R10
147
50   19

TOP PLUNGER
∨∨∨ ∨∨
MATERIAL: 42CrMo4
HARDNESS : 40-42 HRC
QUANTITY: 3 NOS

75   40
640   675
725   600   690   50

BOTTOM PLUNGER, DIE AND HOLDING PLATE

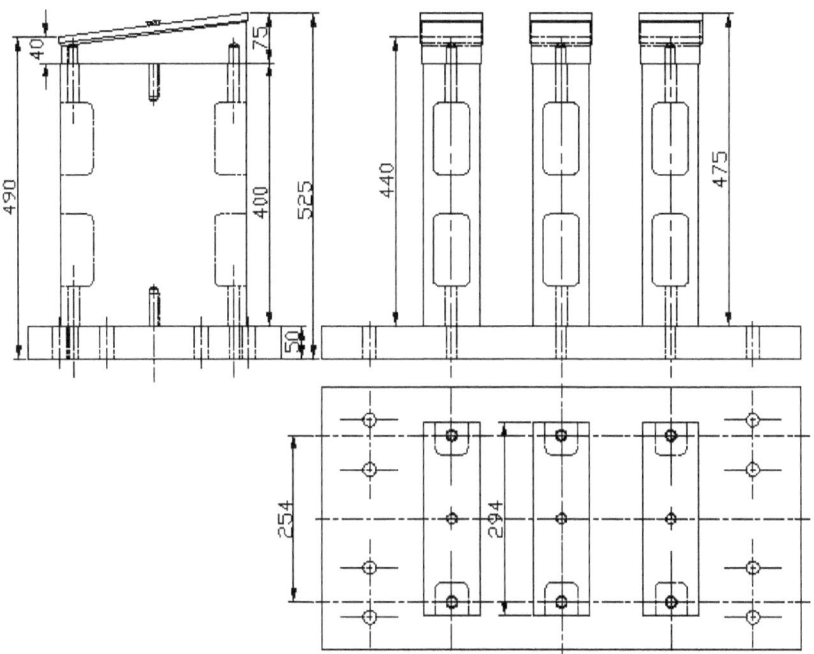

TOP PLUNGER, DIE AND HOLDING PLATE

Liners, packing, lock plates and cover plates assembly:

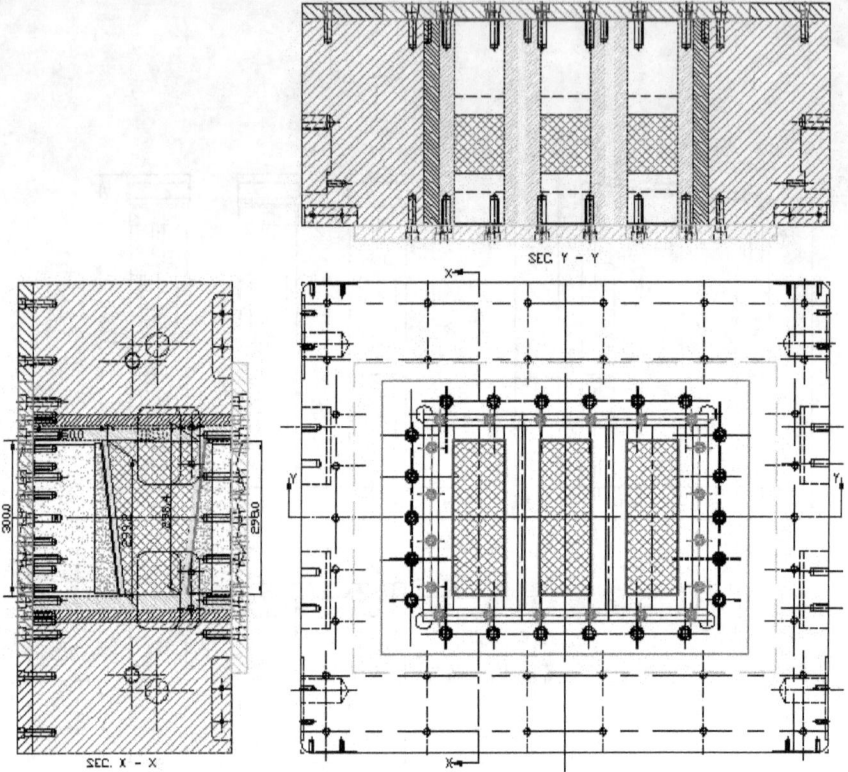

SEC. Y – Y

SEC. X – X

Press configuration for making 300 X 185 / 115 X 100 key brick:

Daylight = 3390, Stroke length of mould sliding table = 1250

Distance from ground level to plunger holding base = 368, Spacer height = 992

Charger height = 340

Bottom plunger height = Plunger holding plate thickness + bottom plunger height + Bottom Die thickness at thicker end = 50 + 600 + 75 = 725

Top plunger height = Plunger holding plate thickness + Top plunger height + Top Die thickness at thicker end = 50 + 400 + 75 = 525.

Distance between top and bottom Dies = Daylight – (Distance of plunger holding base + bottom die height with bottom plunger and holding plate + Top plunger height including Die and holding plate + spacer height) = 3390 – (368 + 725 + 525 + 992) = 780

The press with mould has been configured with charging, pressing and ejection sequence.

Mixture charging in cavities:

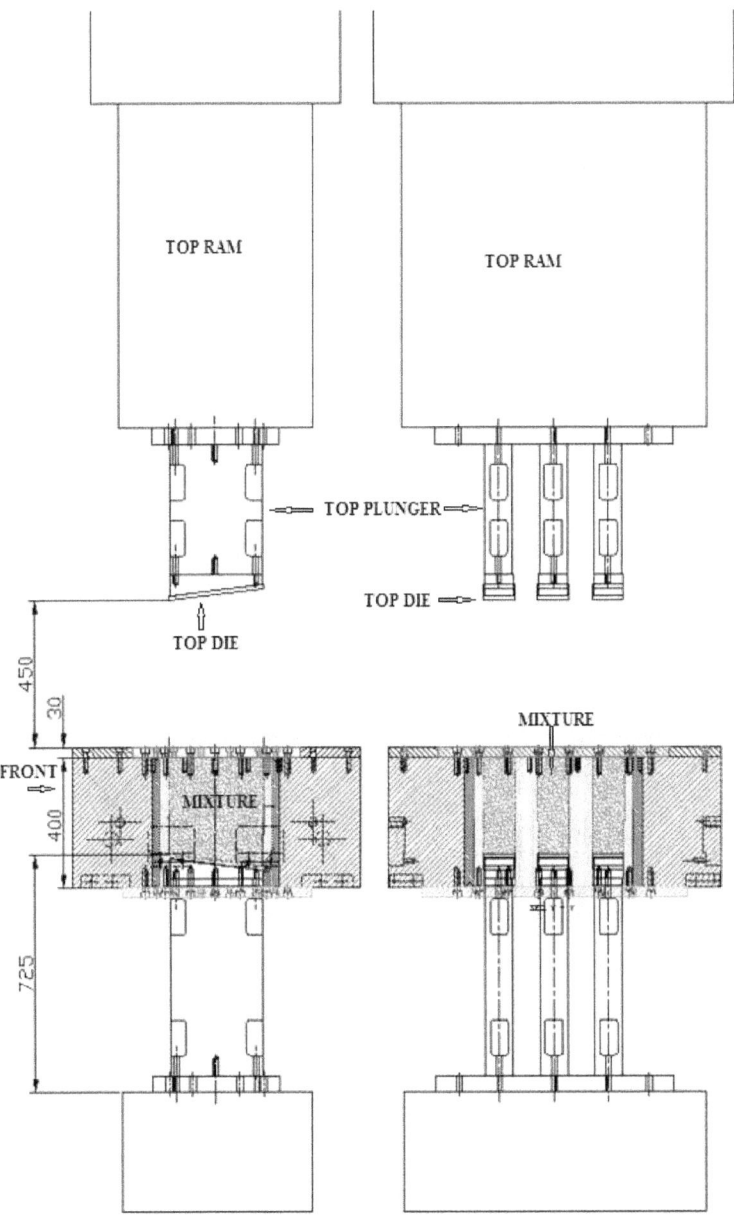

Mixture compressed between dies:

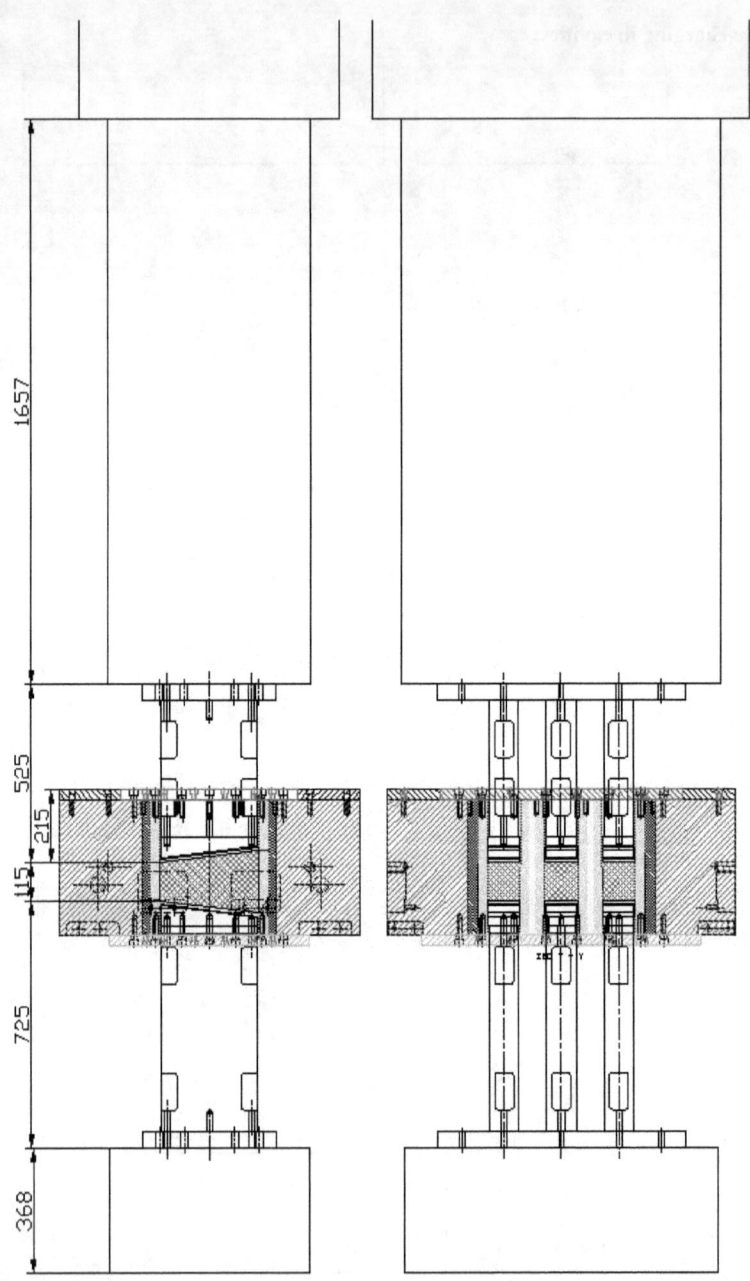

Brick ejected from cavities:

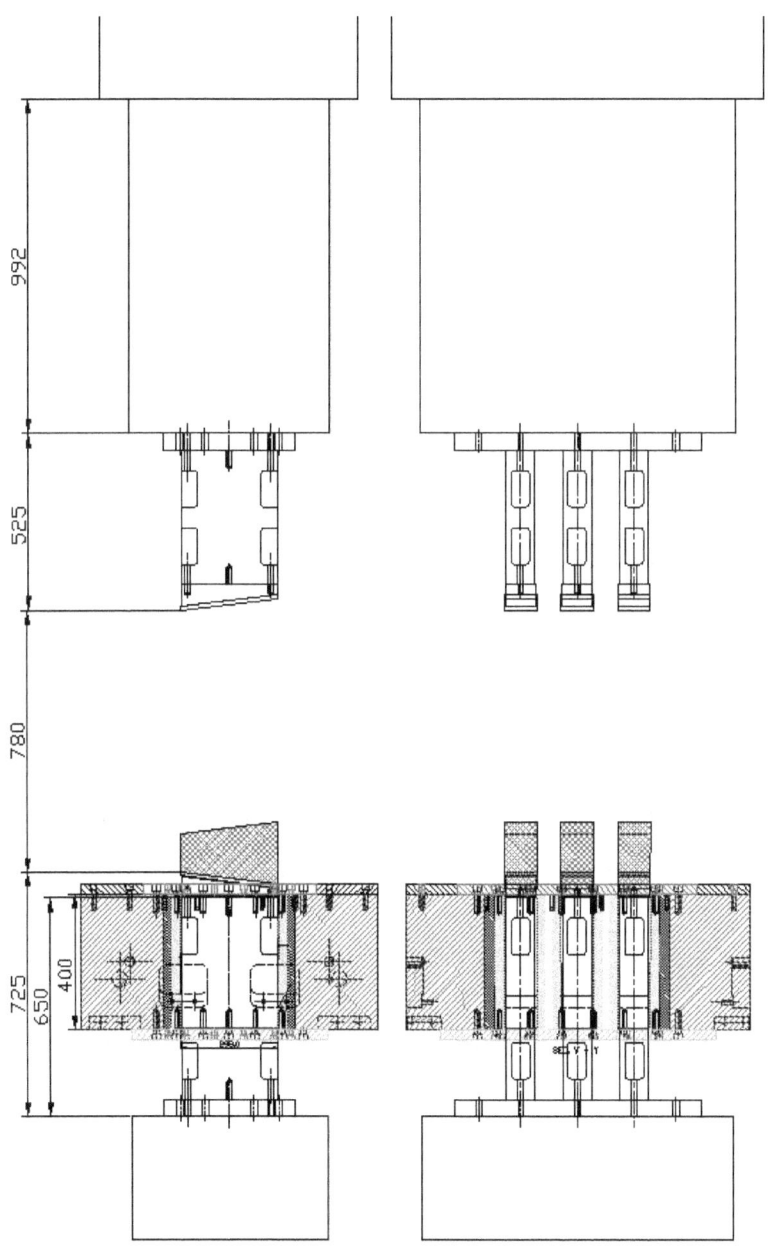

Chapter – 15

Mould design for S.A and E.A Bricks

Relevance of Press in mould design:

- The outer size of mother mould is press dependent. Length and width will vary according to space available in between columns.
- The cavity size and wall thickness will vary according to pressing capacity of press and cavity in mould fixing table of a press.
- Peripheral features on outer walls of mother mould are also press dependent.
- The height of mother mould will vary according to the daylight of press and maximum size of brick thickness that can be pressed.
- No of cavities depends on pressing capacity of press for a particular brick size and specific pressure.
- Length of plungers are also variable depend on working parameters and daylight of press
- Plunger holding plate will also vary from press to press.

Design of moulds for Side Arch and End Arch:

The process of a designing mould for Side Arch and End Arch bricks are same; the only taper on the pressing face will differ. In case of Side Arch taper will be from one side to another side where as in case of End Arch taper will be from one end to another end. The design for End Arch brick has been demonstrated here.

Design mother mould and all accessories required for making End Arch brick of size 450 X 225 X 75/ 65 EA on 1000 Tonnes press having daylight and working parameter as given earlier.

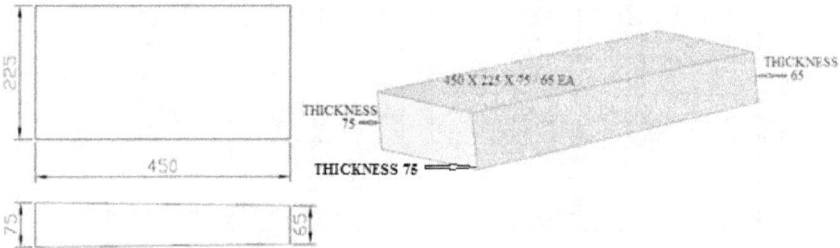

The above-drawn brick is from the group of End Arch bricks serial No 1 of chapter – 13. Let us design mother mould and all necessary accessories for manufacturing on 1000 tonnes hydraulic press with working parameters already mentioned in the previous chapter.

The design of mother mould for any specific press has some permanent feature that cannot be changed. The cavities of mother mould can be changed to certain limitations that have been already described earlier. The height is also changeable to suit the filling depth and maximum thickness of brick that can be moulded. While thinking of lower height of mother mould; the permanent features on four sides should not be disturbed.

The cavity size in this mother mould has been decided to accommodate maximum sizes of End Arch bricks listed in the table of end Arch bricks in chapter -13.

The cavity size for above brick has been decided as 600 X 400 for assembly of other sizes also in the group. The mould height has been decided 250 mm with a similar purpose. The drawing of mother mould with detailed measurement is given below.

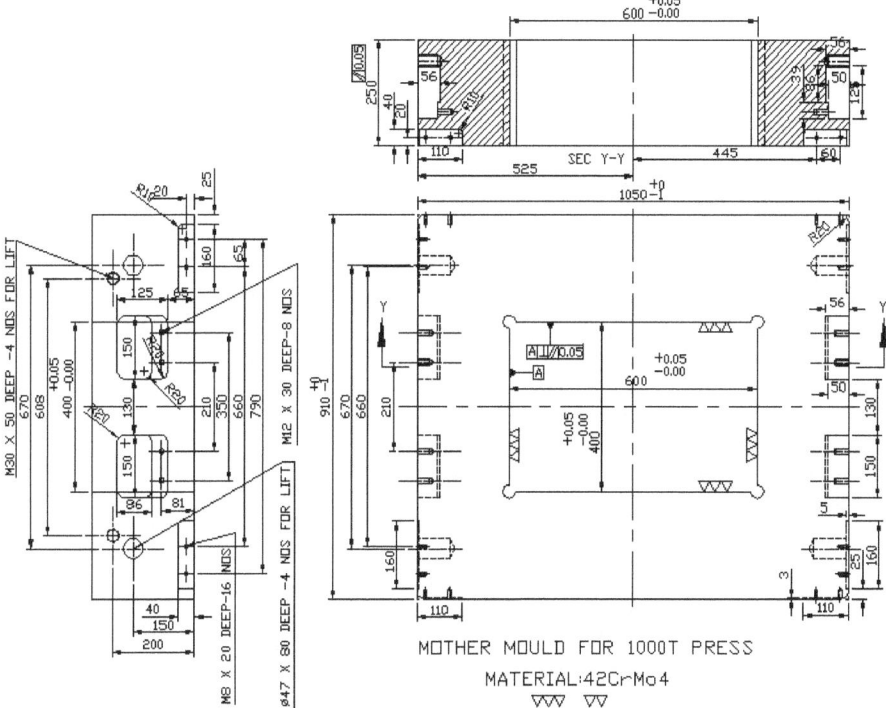

MOTHER MOULD FOR 1000T PRESS
MATERIAL:42CrMo4

Tap holes design on top and bottom face of mother mould:

The design of tap holes is very important in keeping liners and the packing-plates secure in position while compressive force is applied. The location of tap holes will change with a change in cavity size. The centre of holes should be 25 mm away from the end of cavity walls. The size should be strong enough to hold lock plates rigidly.

The all necessary dimensions of tap holes are given in the drawing.

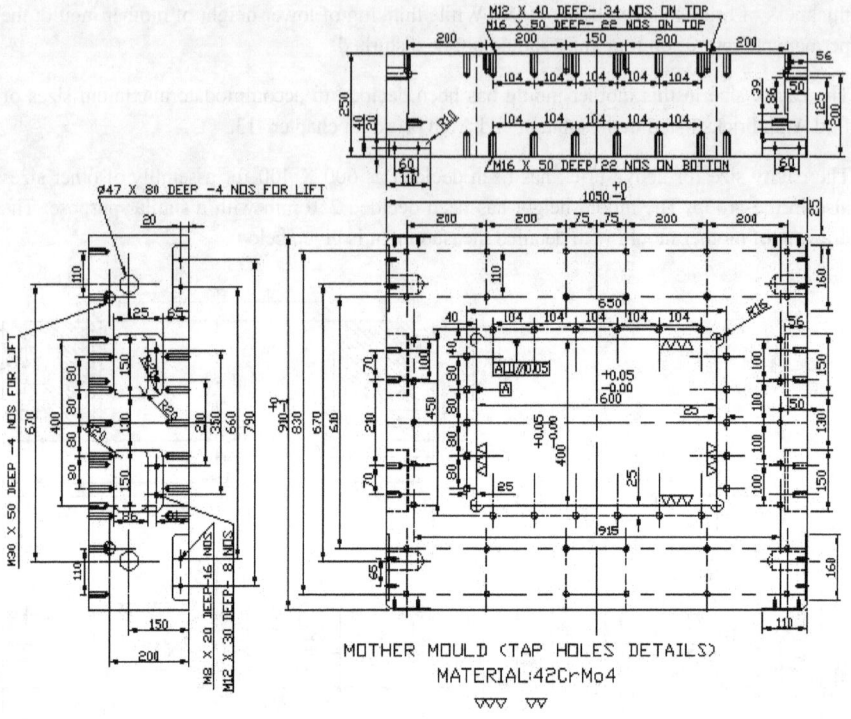

MOTHER MOULD (TAP HOLES DETAILS)
MATERIAL:42CrMo4

Packing design:

The packings are also essential accessories to keep liners rigidly in the mother mould cavity. The length, width, thickness and tap holes details for side and end packing are given in the drawing.

The length of side packing is 1 mm less than cavity length. The packings with 1mm less length will be easy to place inside cavity.

The length of end packing = 400 – (58.5*2) – 0.5 = 282.5 mm. It is also 0.5 mm less to place in between two side packings comfortably. 58.5 is the thickness of side packing.

The thickness of side packing and end packings has been calculated with reference to mould cavity size and liners thickness at the top. The height of packings has been justified 0.5 mm bigger than mould height just to have the better pressure of lock plates with bolts on it from top and bottom. The tolerance and parallelism are also important that have been marked in drawings.

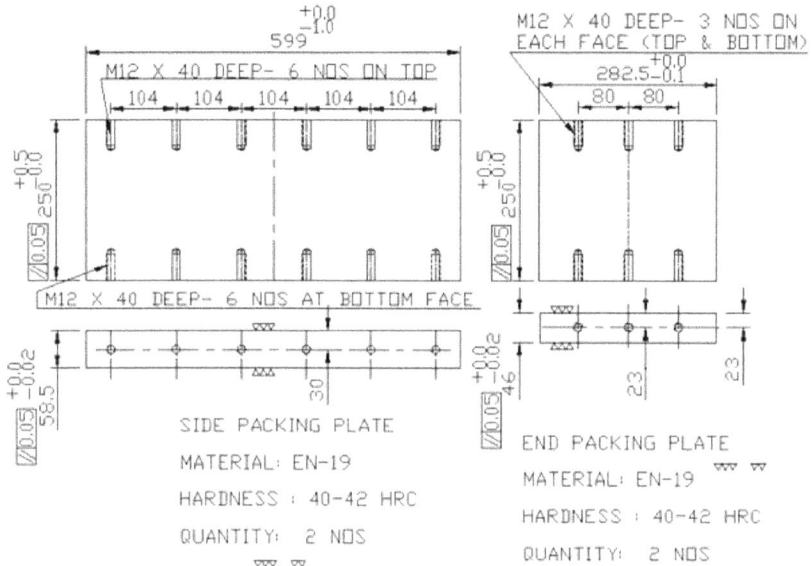

SIDE PACKING PLATE
MATERIAL: EN-19
HARDNESS : 40-42 HRC
QUANTITY: 2 NOS

END PACKING PLATE
MATERIAL: EN-19
HARDNESS : 40-42 HRC
QUANTITY: 2 NOS

Design plan of liners and Dies:

Before designing liners, the tolerance and taper allowed on brick sizes must be considered. The liners sizes have been calculated here on the basis of top cavity size as 450 X 225 and bottom cavity size at a depth of 250 as 448 X 223 mm. this decision has been taken on the basis of tolerance on brick sizes and permissible taper.

- Tolerance ± 1.5 mm
- The taper on ejected surface (surface coming in contact to liners) = 0.8 Maximum.

The top and bottom cavity size and liner face taper will vary with change in tolerance and brick taper.

The draft taper can be reduced to 0.5 mm for closer tolerance and taper.

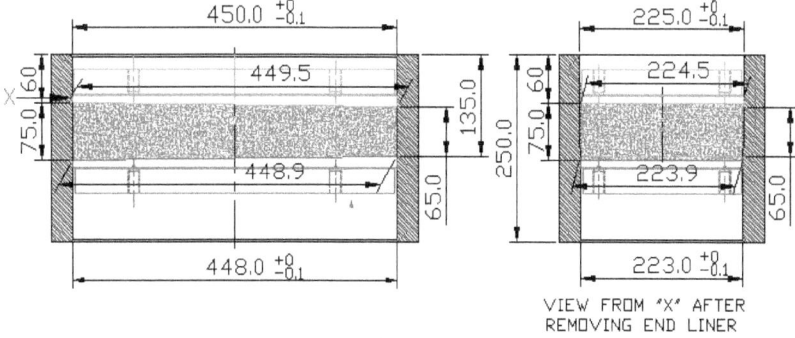

VIEW FROM "X" AFTER
REMOVING END LINER

Liners design:

Side liner: The length of the liner is 0.2 mm less than space in the cavity after setting end packings. It is just to place comfortably in between end packings. The height is in positive side maximum to + 0.5, but must be equal to packing height for having equal pressure on packing and liners. The thickness of liner at the top face is 29 mm and at the bottom face 30 mm. 1 mm differences is face taper to eject brick from the mould cavity. The height of liner must be equal to the height of mother mould and packings.

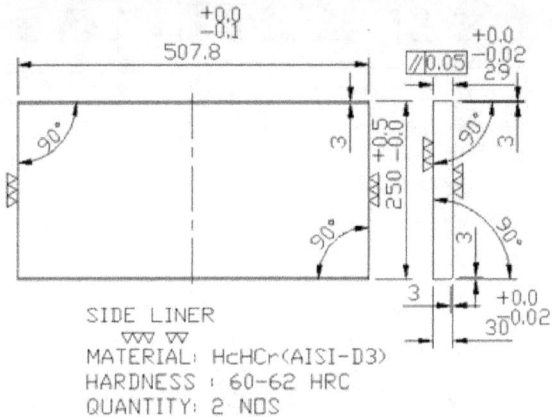

SIDE LINER
MATERIAL: HcHCr(AISI-D3)
HARDNESS : 60-62 HRC
QUANTITY: 2 NOS

End liner: The height of end liner is equal to the height of mother mould and side liners. The tolerance in height must be equal to side liner and packing plates. The width is equal to the brick width at the top and 2 mm less than top at the bottom. It has a matching taper with side liners. The all necessary tolerances and surface finish have been marked in drawings.

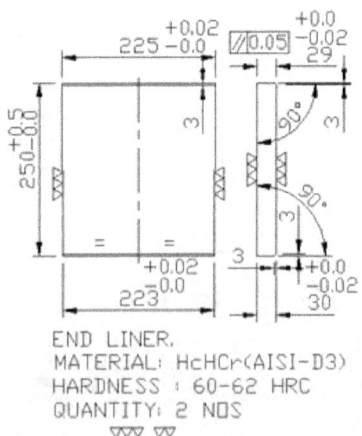

END LINER.
MATERIAL: HcHCr(AISI-D3)
HARDNESS : 60-62 HRC
QUANTITY: 2 NOS

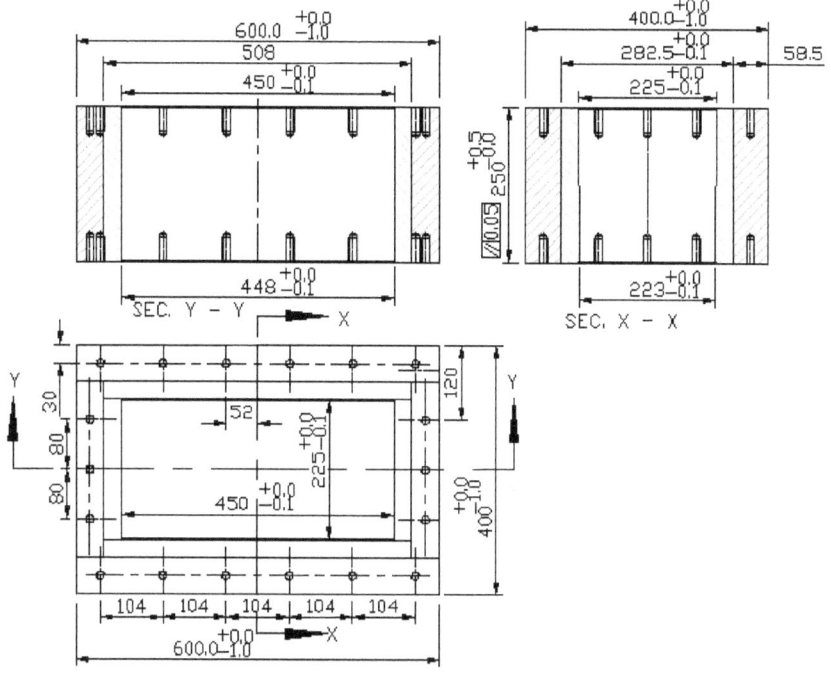

LINERS AND PACKING ASSEMBLY

The liners and packings assembly drawing plotted above have mould cavity size with necessary tolerances. The section X-X and Y-Y have cavity size at top and bottom.

Designs of bottom lock plate

It has counter holes of two different sizes. The bigger size is for fitting lock plate with mother mould and the smaller size is for joining packing plates. The tolerance and parallelism are given in the drawing.

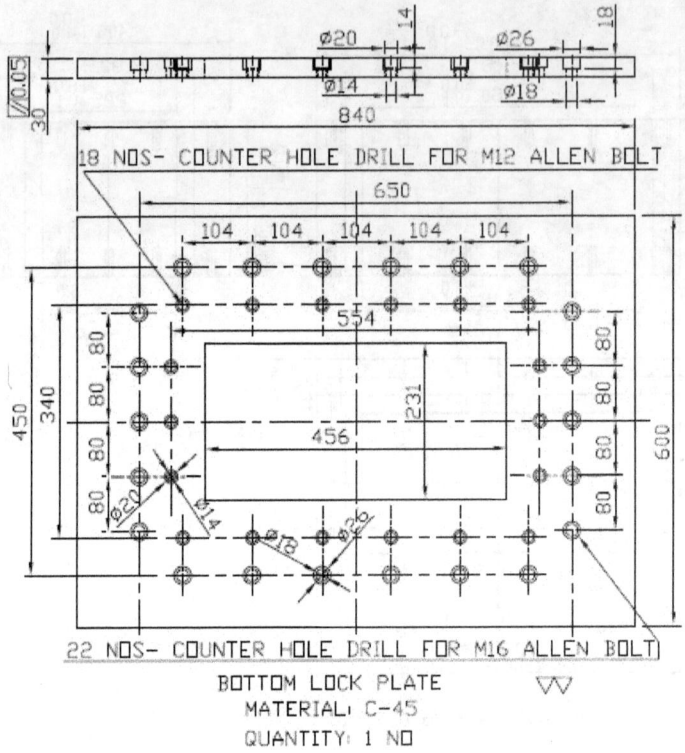

18 NOS- COUNTER HOLE DRILL FOR M12 ALLEN BOLT

22 NOS- COUNTER HOLE DRILL FOR M16 ALLEN BOLT

BOTTOM LOCK PLATE

MATERIAL: C-45

QUANTITY: 1 NO

Top lock plates design:

The top lock plate is also a single piece having matching hole centre with tap holes of mother mould. It has counter holes of two different sizes. The bigger size is for fitting lock plate with mother mould and the smaller size is for joining packing plates. The tolerance and parallelism are given in the drawing. It is smaller in size than bottom lock plate.

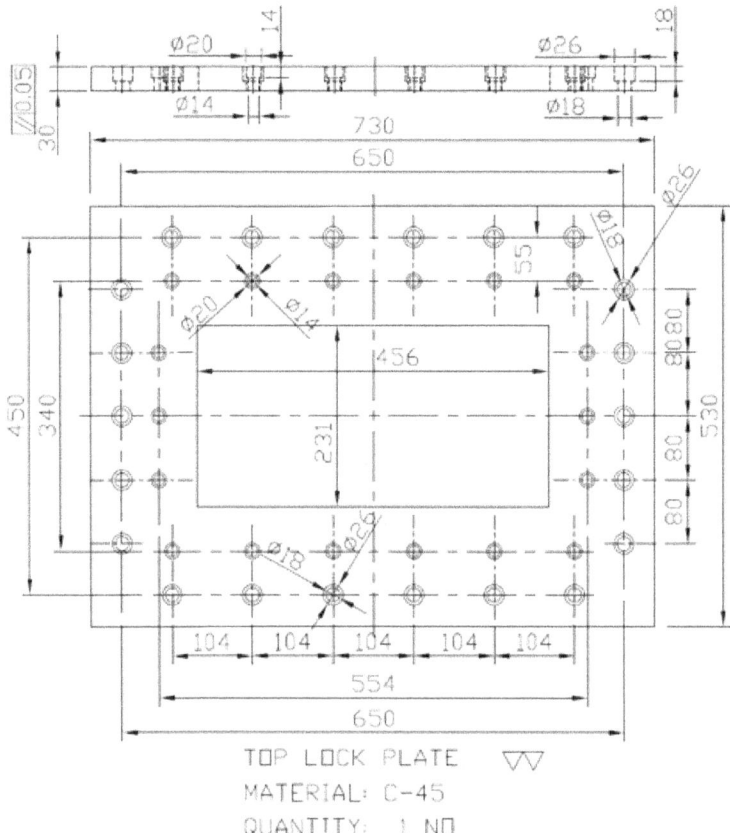

TOP LOCK PLATE ▽▽
MATERIAL: C-45
QUANTITY: 1 NO

Front and back cover plate: These are also part of the top lock plate. The charger has to move to and fro from back to front to charge the mixture in the mould cavity. The top full surface of mould should be smooth and level. These pieces are placed adjacent to top lock plate and fitted with M12 Allen bolts on the top surface of mother mould. The drawing with full detail is plotted below.

There is another way of having a cover plate. In this design, spacers can be fitted on the location of M12 tap holes, on spacers and lock plate top surface again tap holes of smaller size is made to fit a cover plate with counter sink screw bolts. The cover plate will cover the full surface of mother mould top having a cavity in the centre equal to the opening of the lock plate.

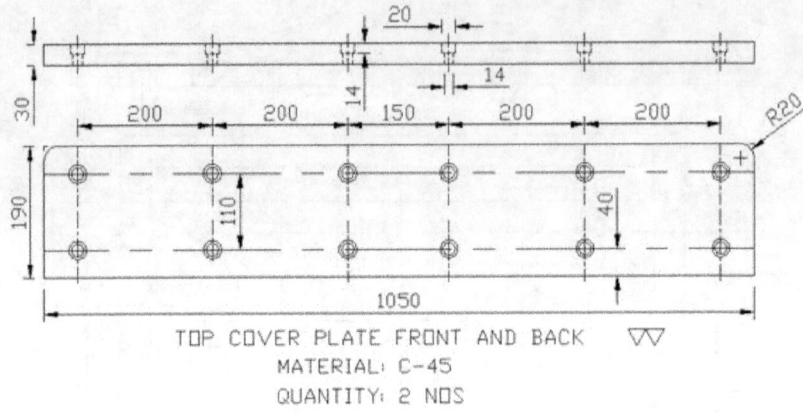

TOP COVER PLATE FRONT AND BACK
MATERIAL: C-45
QUANTITY: 2 NOS

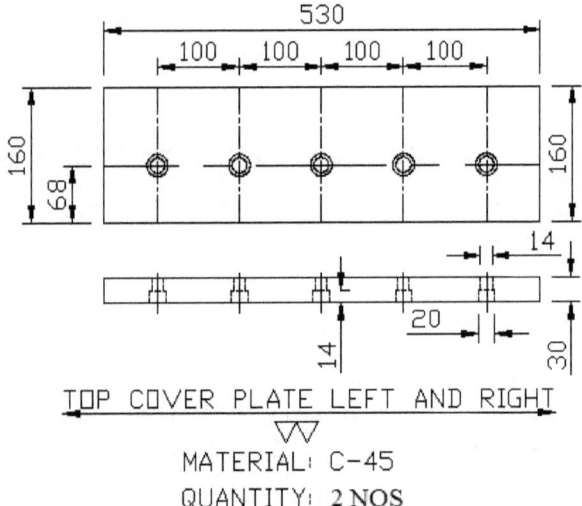

TOP COVER PLATE LEFT AND RIGHT

MATERIAL: C-45
QUANTITY: 2 NOS

Top and bottom die design:

The top and bottom dies have been drafted or designed on the basis of design plan layout. The tap holes have matching location and size with drill holes in plunger pockets.

Top Die: the measurement of length and width of the die will match at filling depth which is 60 mm from the top surface.

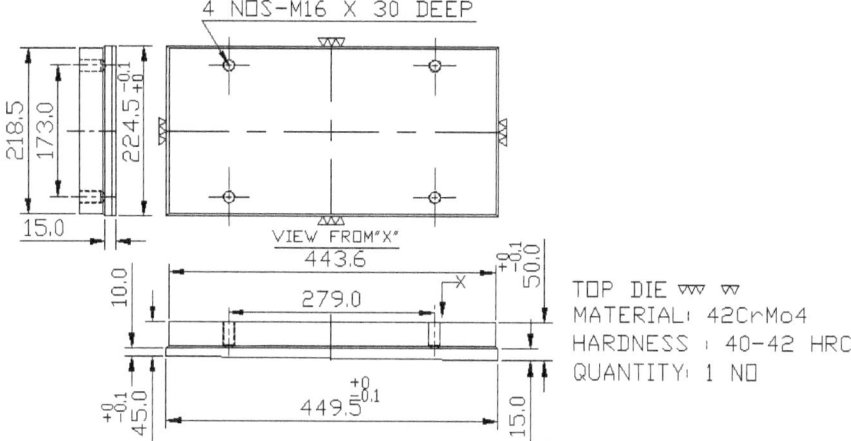

Bottom Die: The shape is same as top die but length and width are smaller by 0.6 mm matching distance at 135 from top

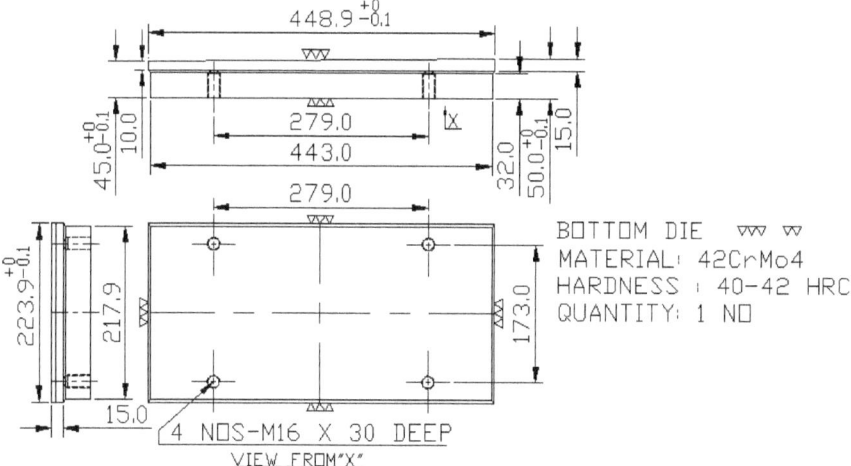

Bottom and Top Plunger design:

The plunger length, pockets, drill holes and outer periphery have been designed as explained earlier. The tolerance and parallelism mentioned here are important and must be maintained for all plungers. The deviation will reflect on bricks.

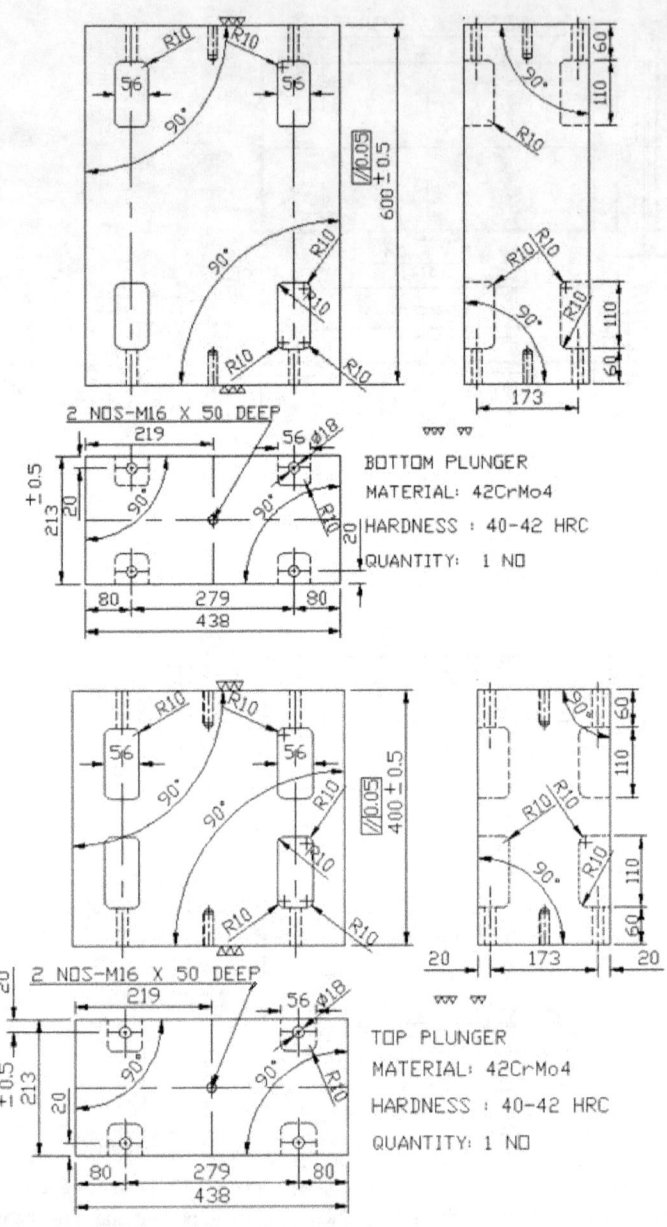

BOTTOM PLUNGER

MATERIAL: 42CrMo4

HARDNESS : 40-42 HRC

QUANTITY: 1 NO

2 NOS-M16 X 50 DEEP

TOP PLUNGER

MATERIAL: 42CrMo4

HARDNESS : 40-42 HRC

QUANTITY: 1 NO

Plunger holding plate:

It is an important accessory of the hydraulic press mould. It connects the plunger and dies to press top ram, bottom ram or bottom plunger fitting base. The outer size (length, width and thickness) must suit the press design to fix at the respective positions. The parallelism is very important for making defect-free brick. The location and size of tap holes must match with drill holes in plunger pocket. The required suitable material and hardness mentioned here must be maintained. The higher grade of material with better hardness may be considered but it should not crack due to a compressive load.

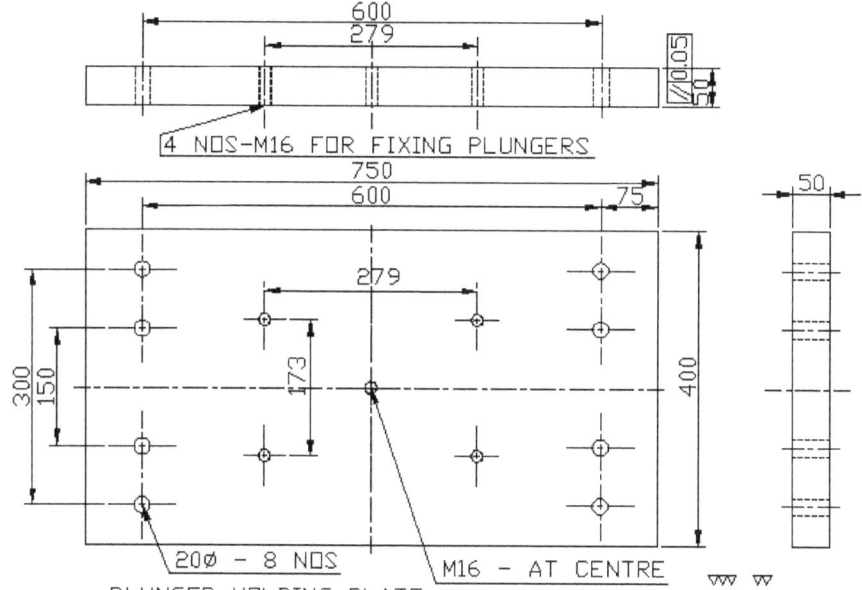

PLUNGER HOLDING PLATE
MATERIAL: 42CrMo4
HARDNESS : 40-42 HRC
QUANTITY: 2 NOS

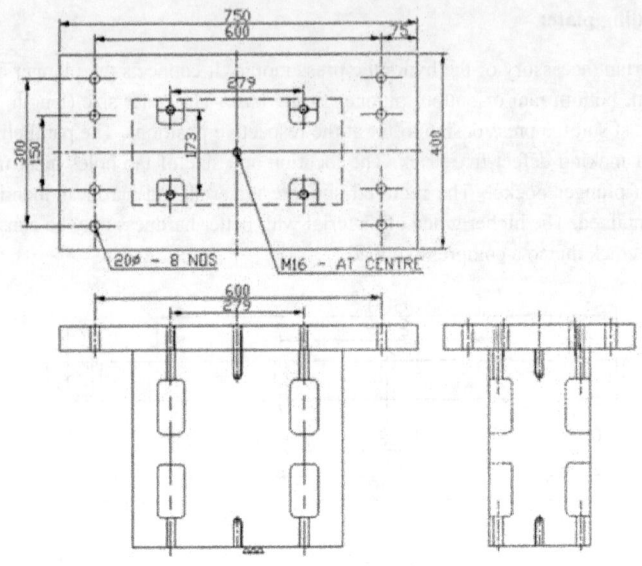

TOP PLUNGER WITH PLUNGER HOLDING PLATE

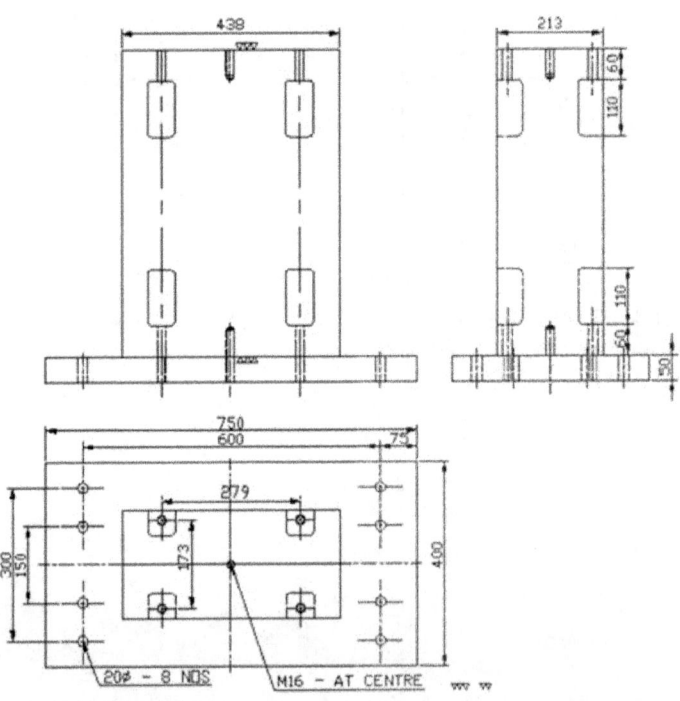

BOTTOM PLUNGER ON PLUNGER HOLDING PLATE

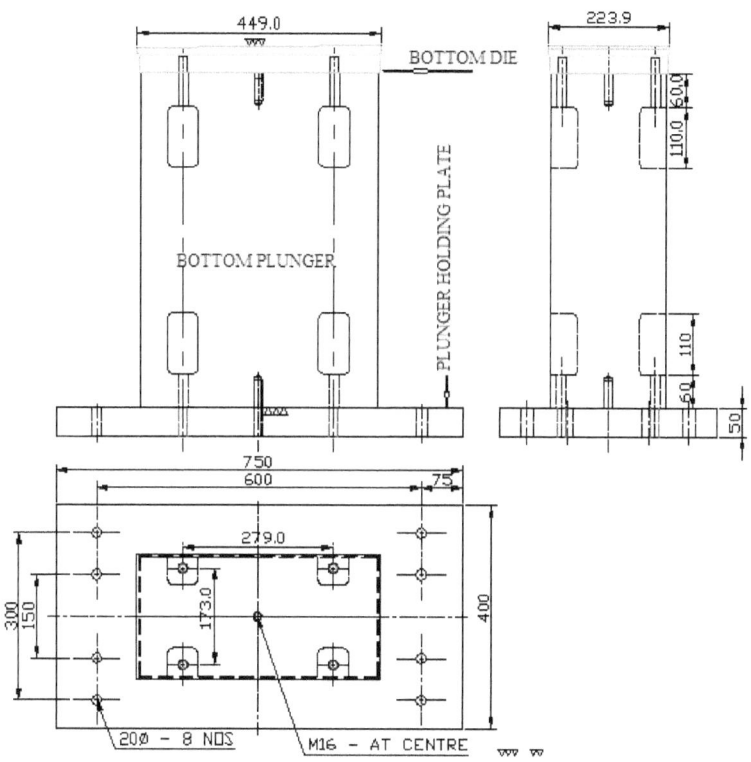

BOTTOM DIE, PLUNGER AND HOLDING PLATE

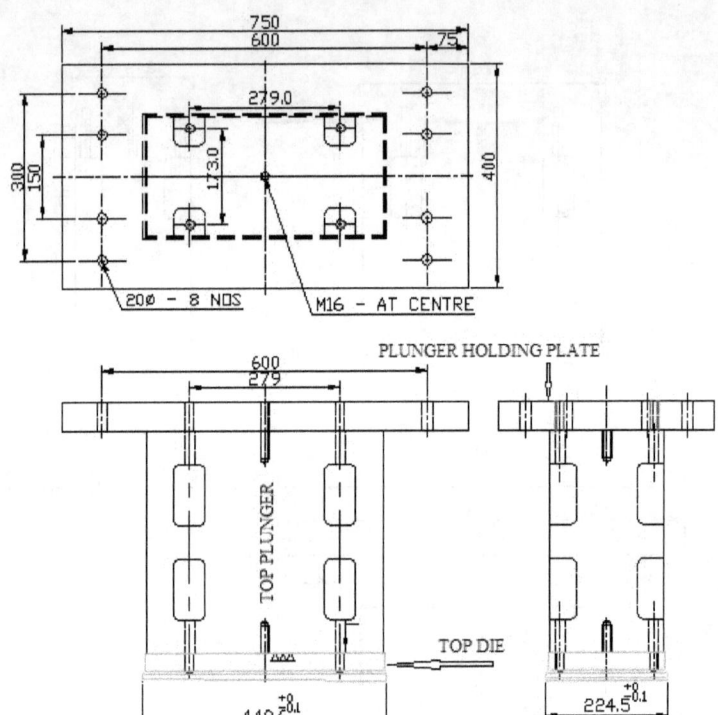

PLUNGER HOLDING PLATE

20∅ — 8 NOS

M16 — AT CENTRE

TOP PLUNGER

TOP DIE

TOP DIE AND PLUNGER ON HOLDING PLATE

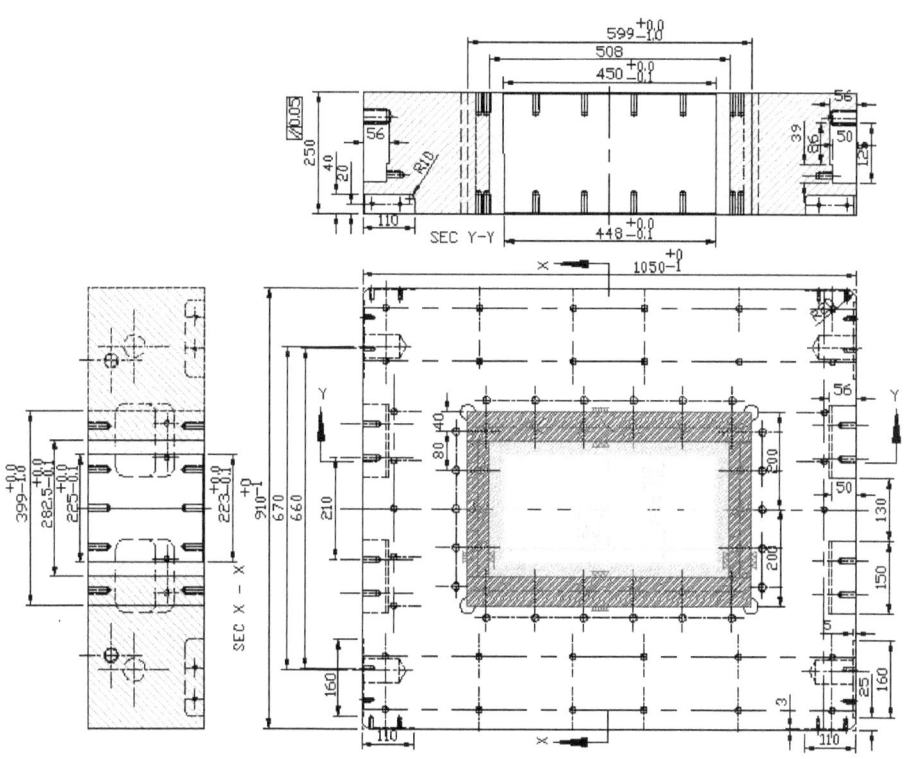

LINERS AND PACKING ASSEMBLY IN MOTHER MOULD

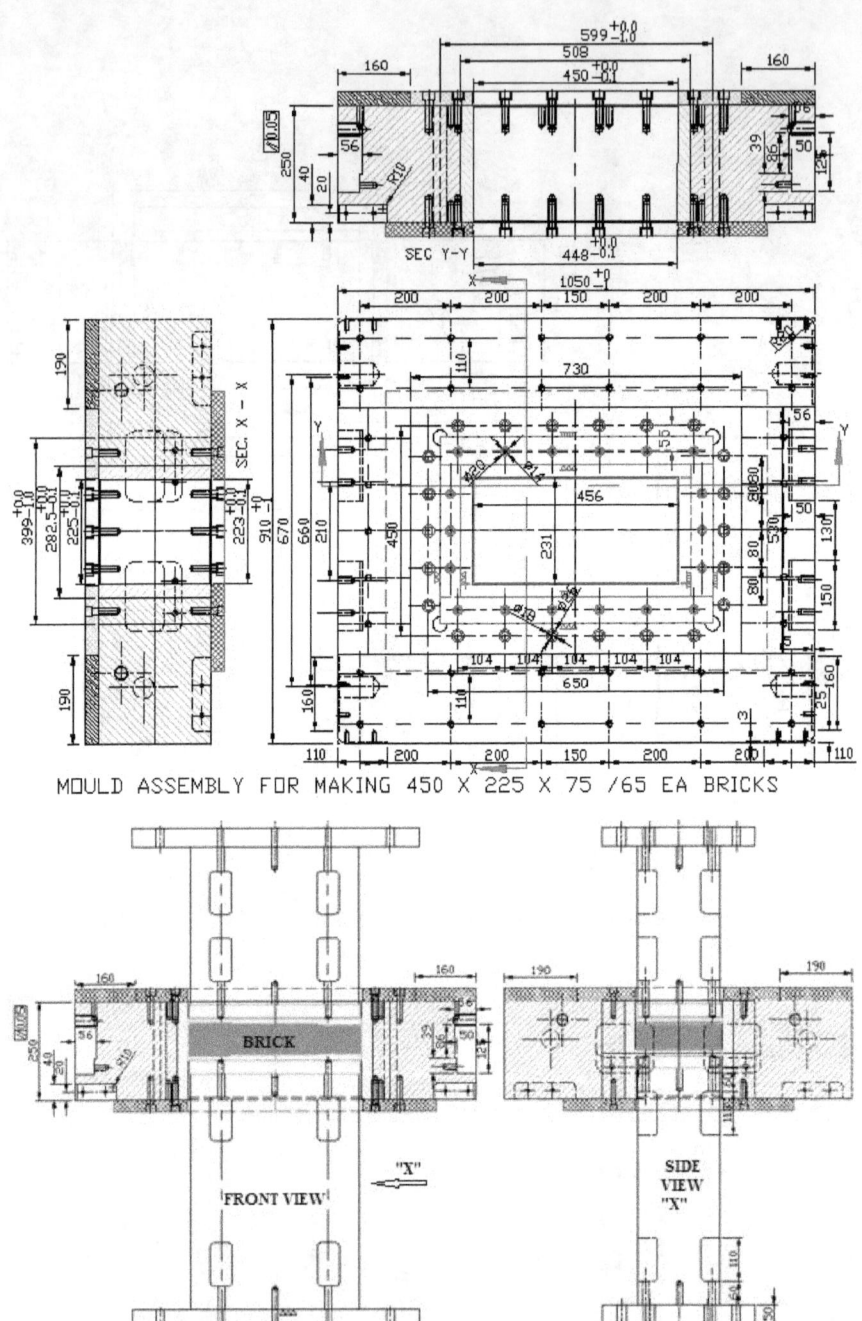

MOULD ASSEMBLY FOR MAKING 450 X 225 X 75 /65 EA BRICKS

BRICK

"X"

FRONT VIEW

SIDE
VIEW
"X"

Chapter – 16

Mould design for silica brick

Note: The shape and sizes of the brick given here are not for application. It has been developed and designed just to explain the process of mould design and to educate the learners.

Let us design a mould for a coke-oven brick that will expand 5% on firing. The simple way of visualisation will illustrate you to deduct 5 from 100 to make a mould with 95 measurements. Working in this way will land you to the wrong path. The measurement of 95 mm in brick will not be 100 mm after firing. 5% means 100 mm will expand to 105 after firing.

If the measurement of brick is 105 mm you have to deduct 5 mm to make 100mm green brick size. The actual deduction from given brick size will be (5*100/105) = 4.762%. Deduct this amount from given brick sizes to design the mould. If brick size is 100 then, 100 – 4.762 = 95.238 mm will be green brick size. It will expand by 5% to have 100 mm on firing.

The layout with – 5 % and – 4.762 % have been plotted below; compare the difference.

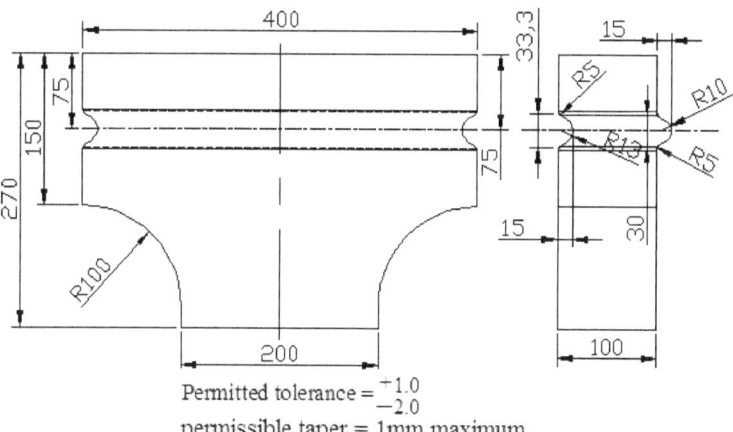

Permitted tolerance $= {}^{+1.0}_{-2.0}$

permissible taper = 1mm maximum

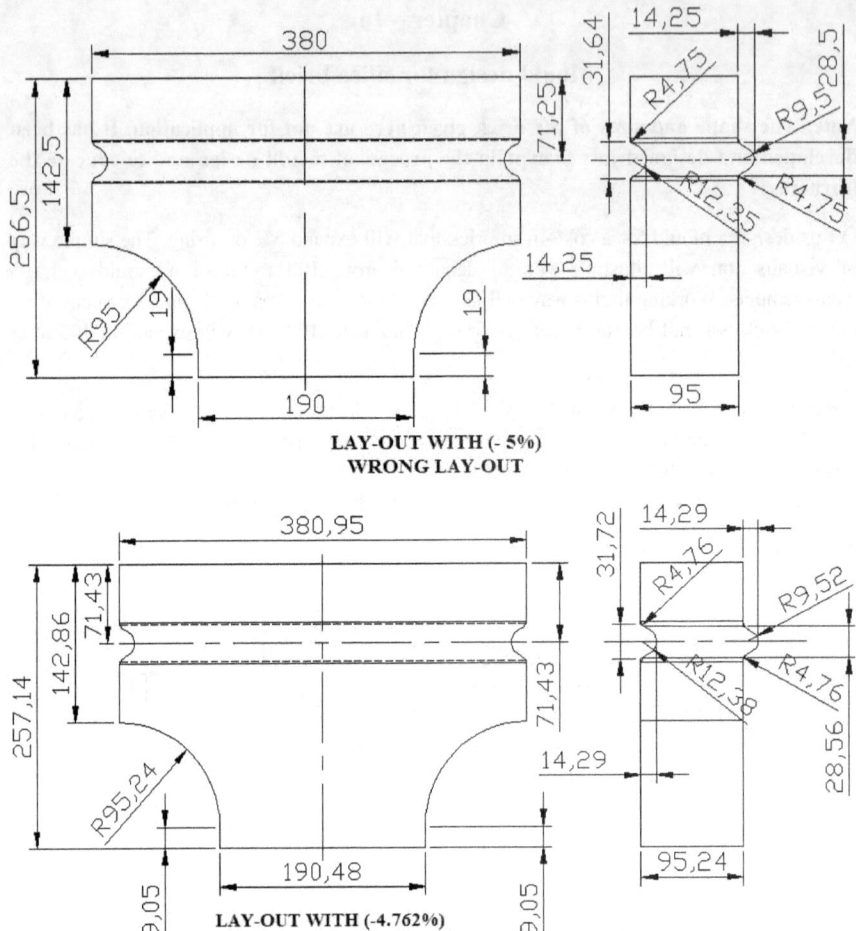

**LAY-OUT WITH (- 5%)
WRONG LAY-OUT**

**LAY-OUT WITH (-4.762%)
CORRECT LAY- OUT**

The mould accessories for this shape will be

- Mother mould
- Sideliners
- End liners
- Loose pieces
- Top die
- Bottom die
- Top plunger
- Bottom plunger
- Top plunger holding plate
- Bottom plunger holding plate

- Bottom lock plate
- Top lock plate
- Cover plate

Mother mould: To illustrate the process 1000 tonnes capacity press with top pressing and mould sliding table has been considered. The working parameter and process of mother mould design, the decision on no of the cavity have been explained earlier. The one cavity mould design on this press with 1.0 or 0.8 tonne/ per sq.cm specific pressure is possible as the area of die face will suit for one cavity only. The mother mould with 500 X 450 cavity size and 250 heights will be suitable for moulding this brick (See Chapter – 13 side arch brick serial no 7 and 8).

The design of mother mould with cavity size and other details are given here.

The cavity size in mother mould can change according to the formation of the group, brick sizes and no of cavities. It has been explained in the previous chapters.

The number of tap holes, location and centre to centre distance also may change according to length and width of the cavity. The radius of circular undercut at four corners of the cavity is 16mm. Its location will also change with a change in cavity size. The tap hole at corners for fitting lock plates should not be very close to it.

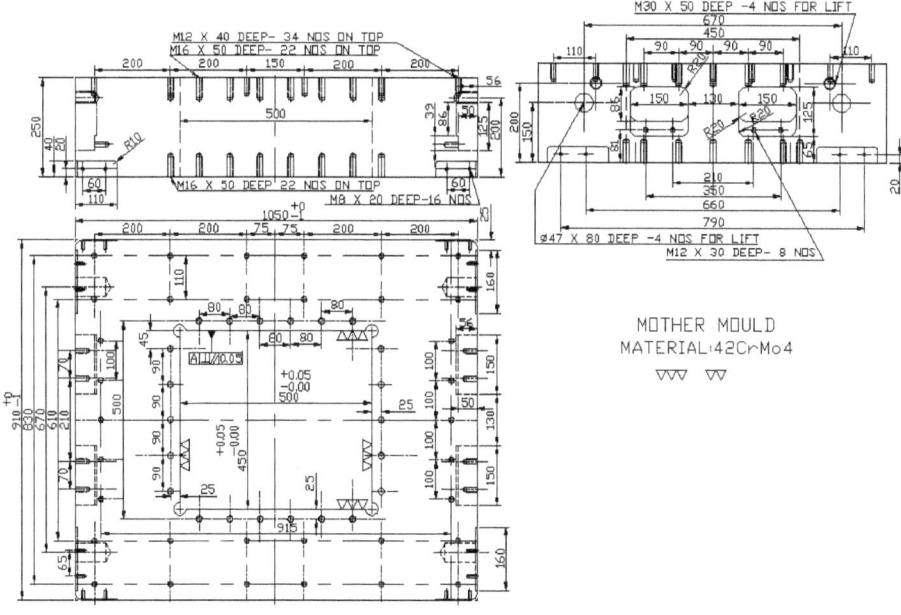

MOTHER MOULD
MATERIAL:42CrMo4

Mother mould's enlarged views of front, plan and side:

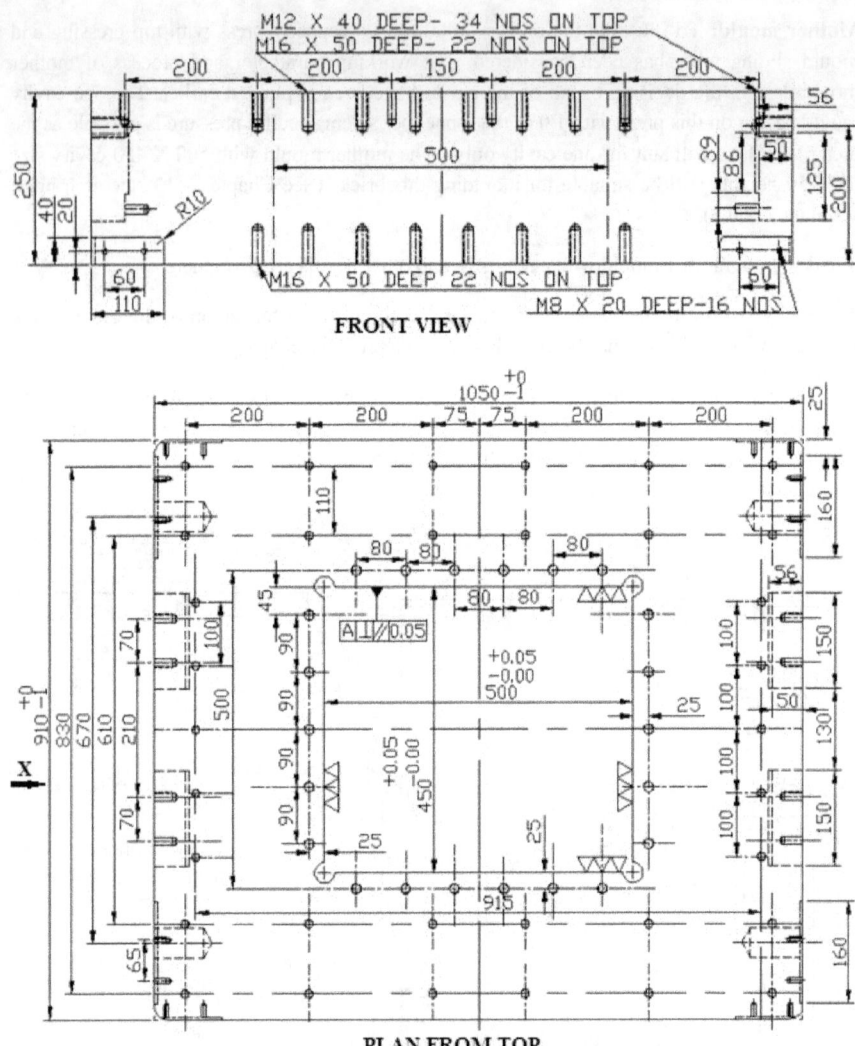

FRONT VIEW

PLAN FROM TOP

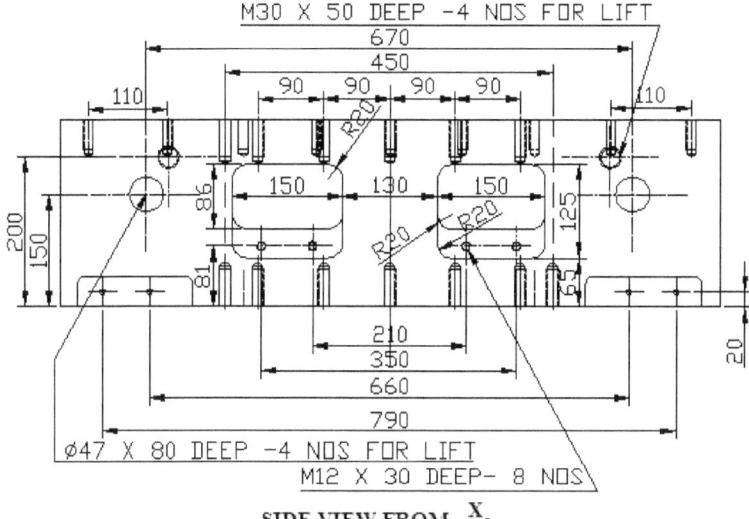

SIDE VIEW FROM X

Overall plan for liners and dies design:

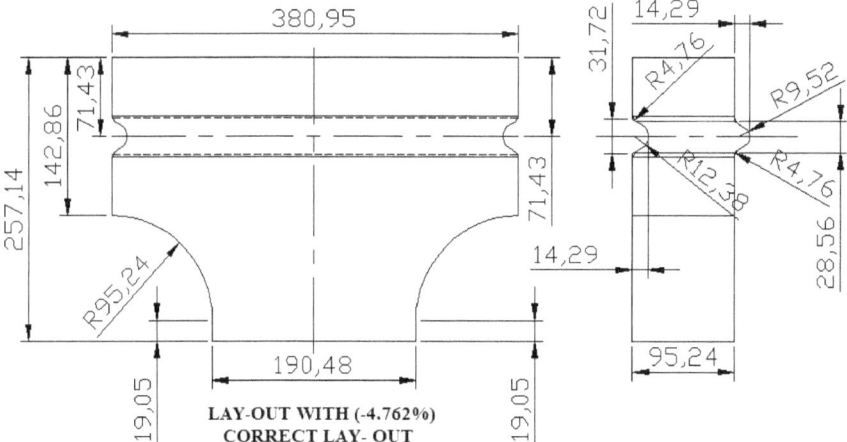

LAY-OUT WITH (-4.762%)
CORRECT LAY- OUT

Consider the layout with expansion allowance as plotted above and make an overall plan to design liners and dies before making individual drawings.

Benefit of overall plan:

- Will eliminate the possibility of mistakes.
- Will provide visualisation of correct shape of side-liners, end liners, loose pieces and dies.
- You will be able to provide face taper on liners and loose pieces

- You will be able to decide required size of mould cavity length and width at top and bottom face
- You will get size of end liner at top and bottom face
- You will get brick sizes at top face and bottom face matching with permissible tolerance.
- You will get a taper in bricks within the permissible range.

The overall design plan to develop the individual drawing has been plotted with nomenclatures.

Option – 1

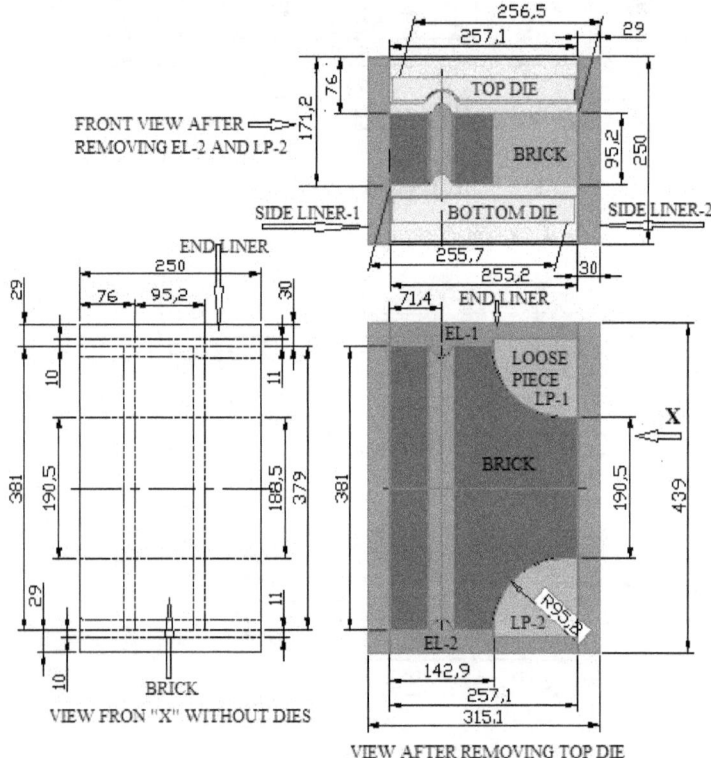

The loose piece and end liners can be designed as drawn in option – 1 if a machine shop can make matching taper of loose pieces in end liners. The end liner will have counterbore drill and loose piece matching tap holes to join together. The assembly will need a bit more care due to unbalanced contact surfaces on both end face with side liners.

Option – 2

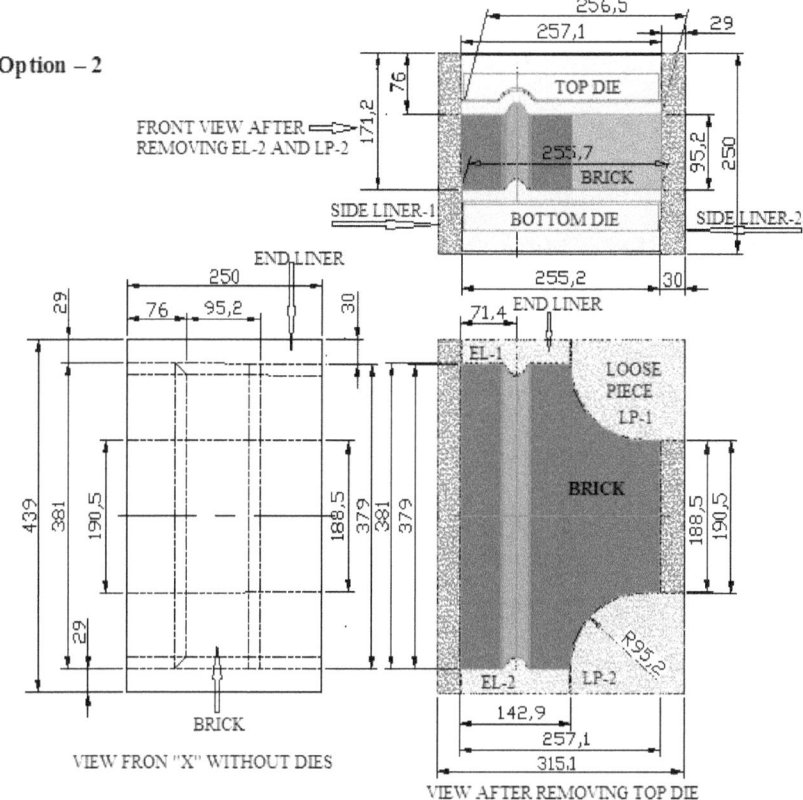

FRONT VIEW AFTER
REMOVING EL-2 AND LP-2

VIEW FRON "X" WITHOUT DIES

VIEW AFTER REMOVING TOP DIE

Individual components of option – 2 have been developed and plotted below from overall design plan.

Formation of mould cavity in mother mould:

The length of mould cavity decided to place across the width of mother mould cavity. The width of mould cavity will be along the length and tight fitted with end packings on both sides.

Benefits of putting cavity across length of mother mould:

- Gripper has better-balanced hold on brick
- Mixture feeding in cavity by charger will be uniform

Size of side and end packings:

Side packing thickness = (450 – 439) / 2 = 5.5 mm

End packing thickness = (500 – 315.1) / 2 = 92.45 mm

Size of side packing = 500 X 250 X 5, 5

The size of end packing = 439 X 250 X 92.5, the end packing will have tap holes on top and bottom face to fix with lock plates. The tap holes are generally placed in the centre of thickness.

Sideliner – 1:

The material of liners may be selected according to requirement. **Make a tap hole in the centre of the top surface for the handling purpose.**

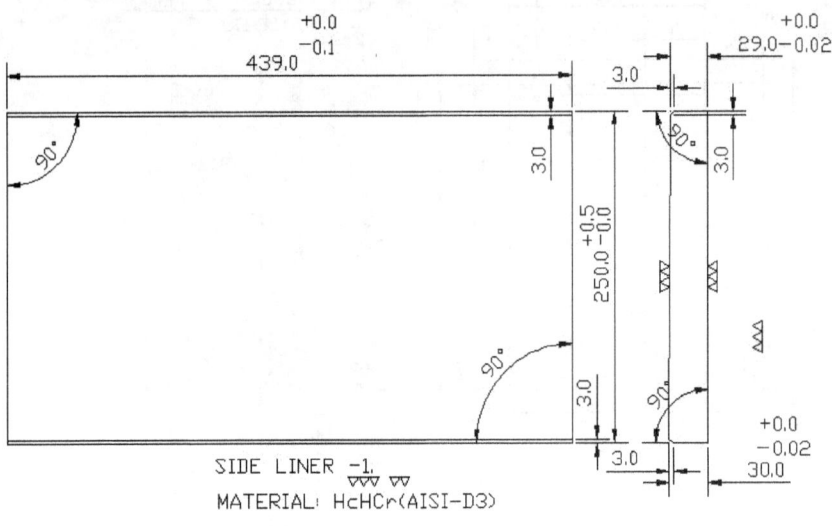

SIDE LINER −1,
MATERIAL: HcHCr(AISI−D3)
HARDNESS : 60−62 HRC
QUANTITY: 1 NO

Side liner – 2 : Generally the side liner do not have a taper on end face but in this case, as end face will be in contact with loose pieces on two ends, matching tapers have been provided. **Make a suitable tape hole in the centre of the top surface for easy handling.**

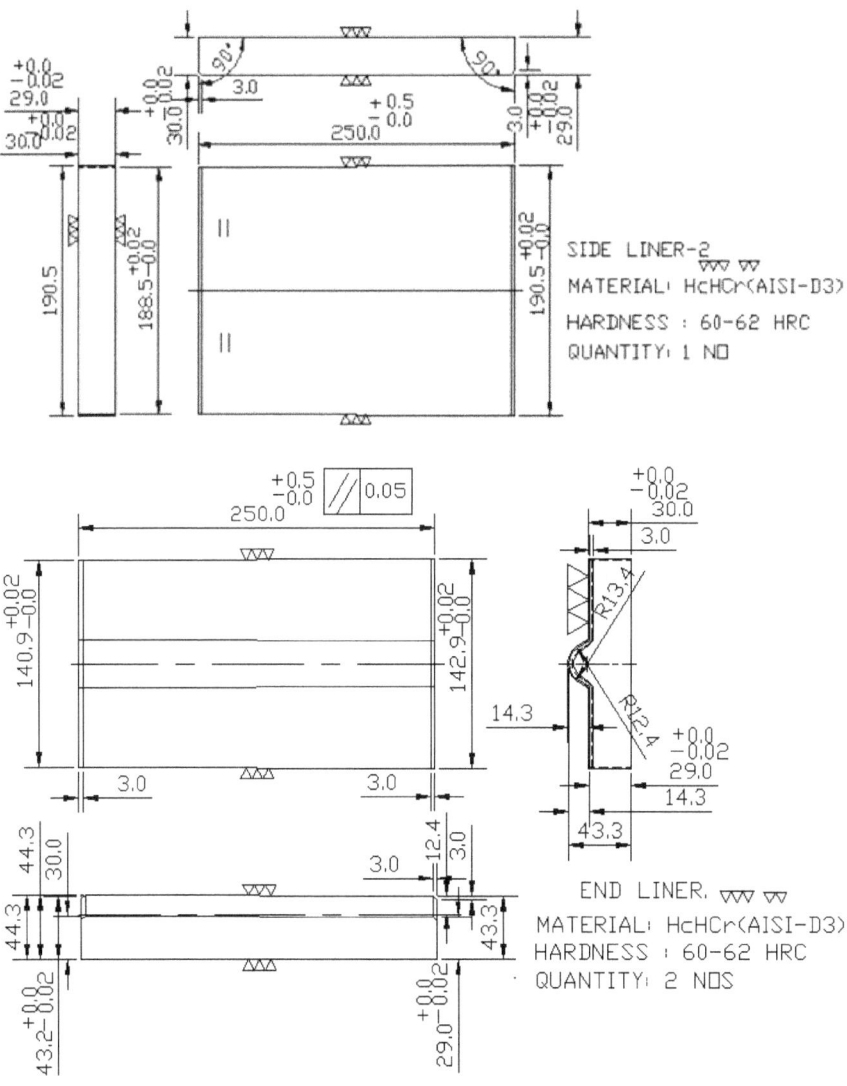

SIDE LINER-2
MATERIAL: HcHCr(AISI-D3)
HARDNESS : 60-62 HRC
QUANTITY: 1 NO

END LINER.
MATERIAL: HcHCr(AISI-D3)
HARDNESS : 60-62 HRC
QUANTITY: 2 NOS

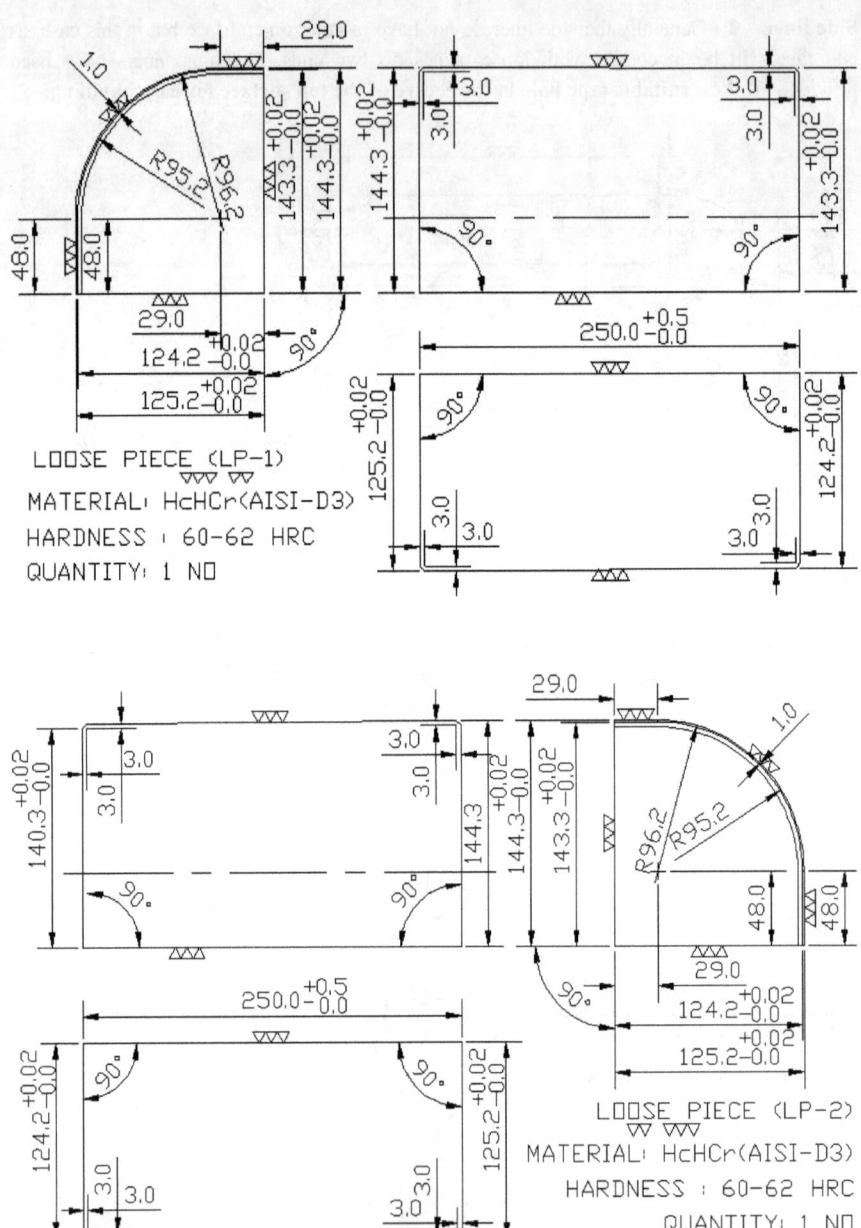

LOOSE PIECE (LP-1)
MATERIAL: HcHCr(AISI-D3)
HARDNESS : 60-62 HRC
QUANTITY: 1 NO

LOOSE PIECE (LP-2)
MATERIAL: HcHCr(AISI-D3)
HARDNESS : 60-62 HRC
QUANTITY: 1 NO

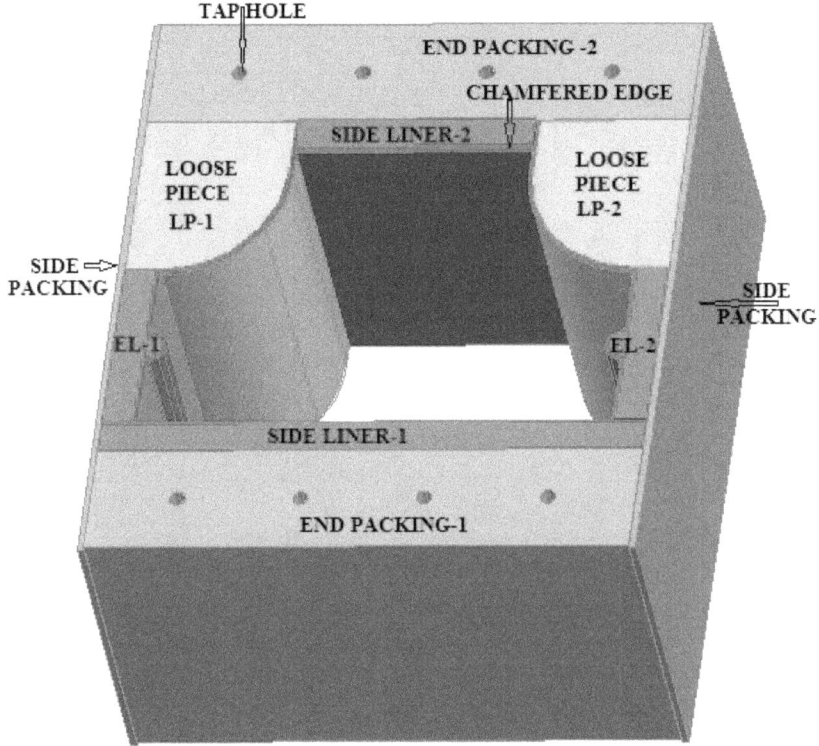

Design and manufacturing of Loose piece – 1 and – 2:

There is the possibility of making two pieces of same size and shape to assemble in two opposite sides of side liner – 2 as it seems to be symmetrical in shape. Study drawing carefully, there is a straight portion on two sides of loose piece – 1 that are coming in contacts with end liner – 1 and side liner – 2. The size of straight length is 48 mm on the side that is coming in contact with side liner -2. The size of straight length that comes in contact with end liner-1 is 29 mm. Similar is the case with loose piece – 2; 48 mm straight portion comes in contact with side liner – 2 and 19 mm in contact with end liner-2.

Two pieces left and right hand must be made to get correct assembly. It may happen in another shape of mould design also. As such it is essential to study the brick drawing thoroughly before mould design and manufacturing accessories.

Design of End liner -1 and End liner-2:

These two liners are identical in shape. 2 pieces of same shape and size can be made to assemble in two ends as shown in the drawing.

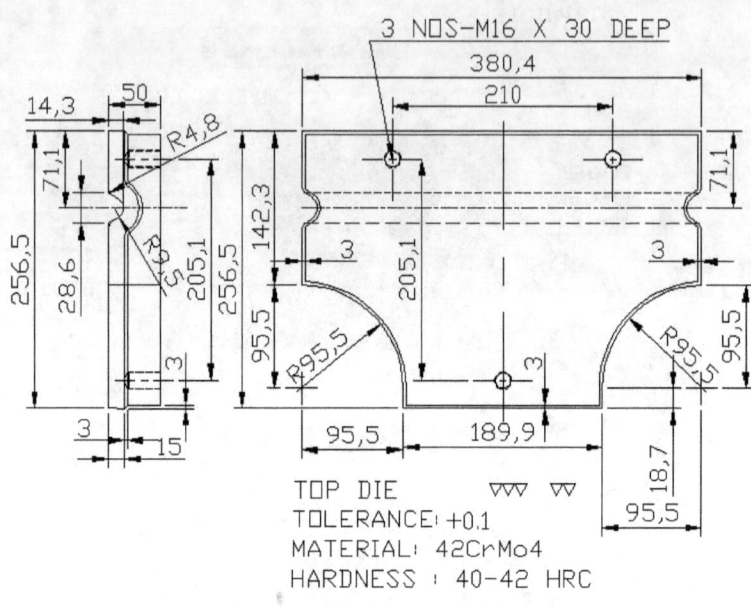

3 NOS-M16 X 30 DEEP

TOP DIE ∇∇∇ ∇∇
TOLERANCE: +0.1
MATERIAL: 42CrMo4
HARDNESS : 40-42 HRC

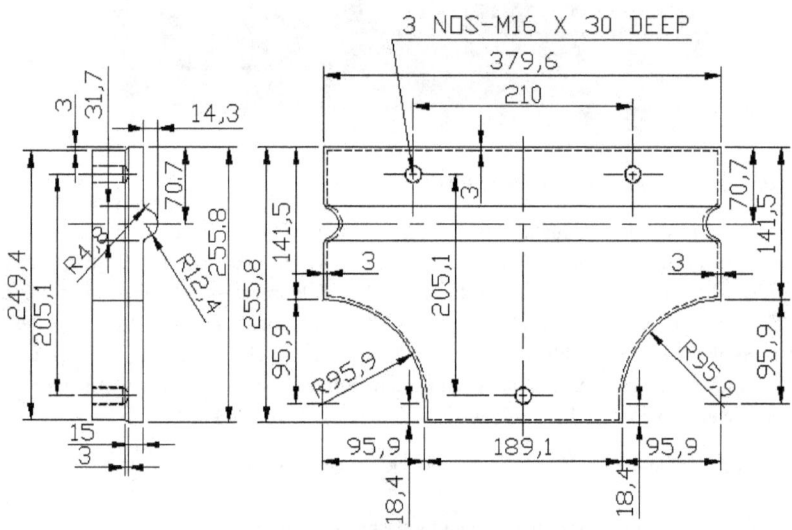

3 NOS-M16 X 30 DEEP

BOTTOM DIE ∇∇∇ ∇∇
TOLERANCE: +0.1
MATERIAL: 42CrMo4
HARDNESS : 40-42 HRC

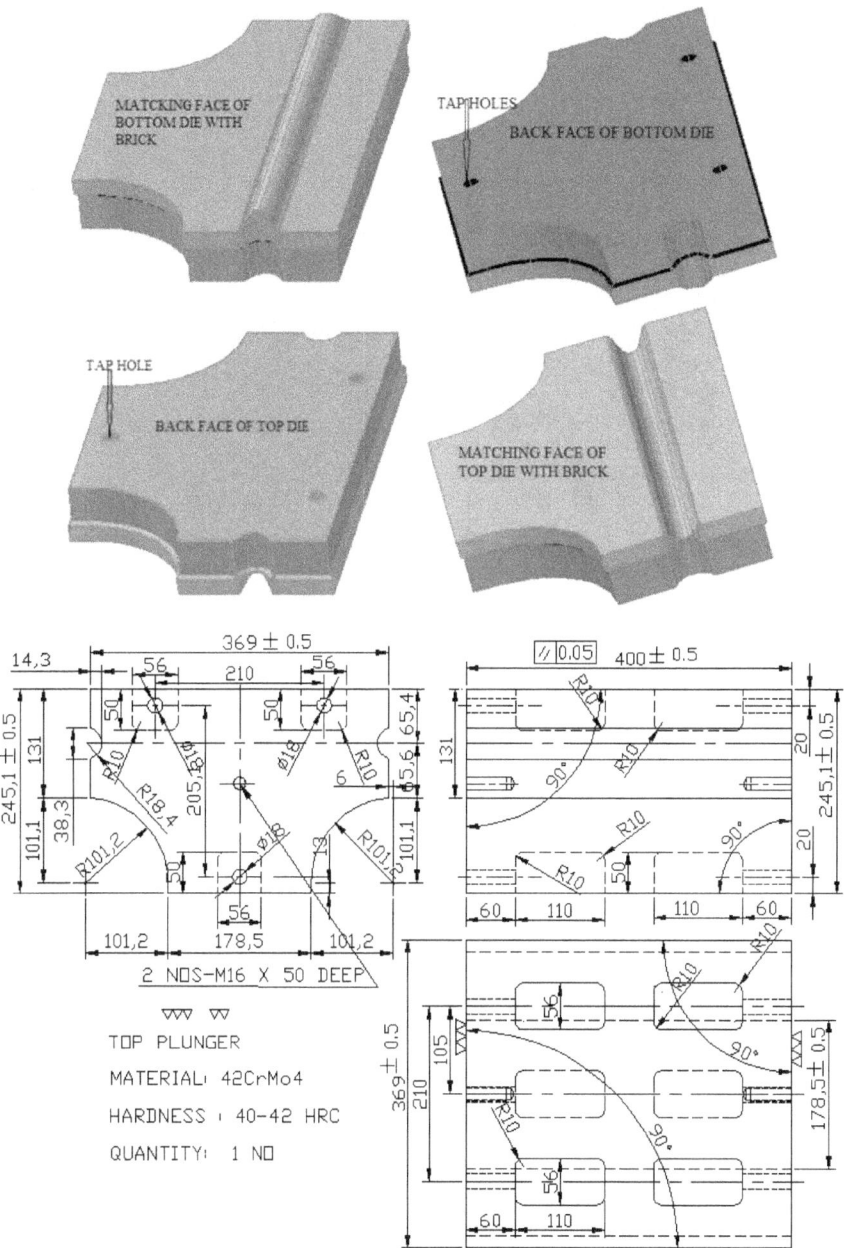

Top plunger 3D view: The top plunger holding plate will be attached with Allen bolt to face written "TOP" and top die to the bottom face. The right side view has been rotated to show pockets and drill holes.

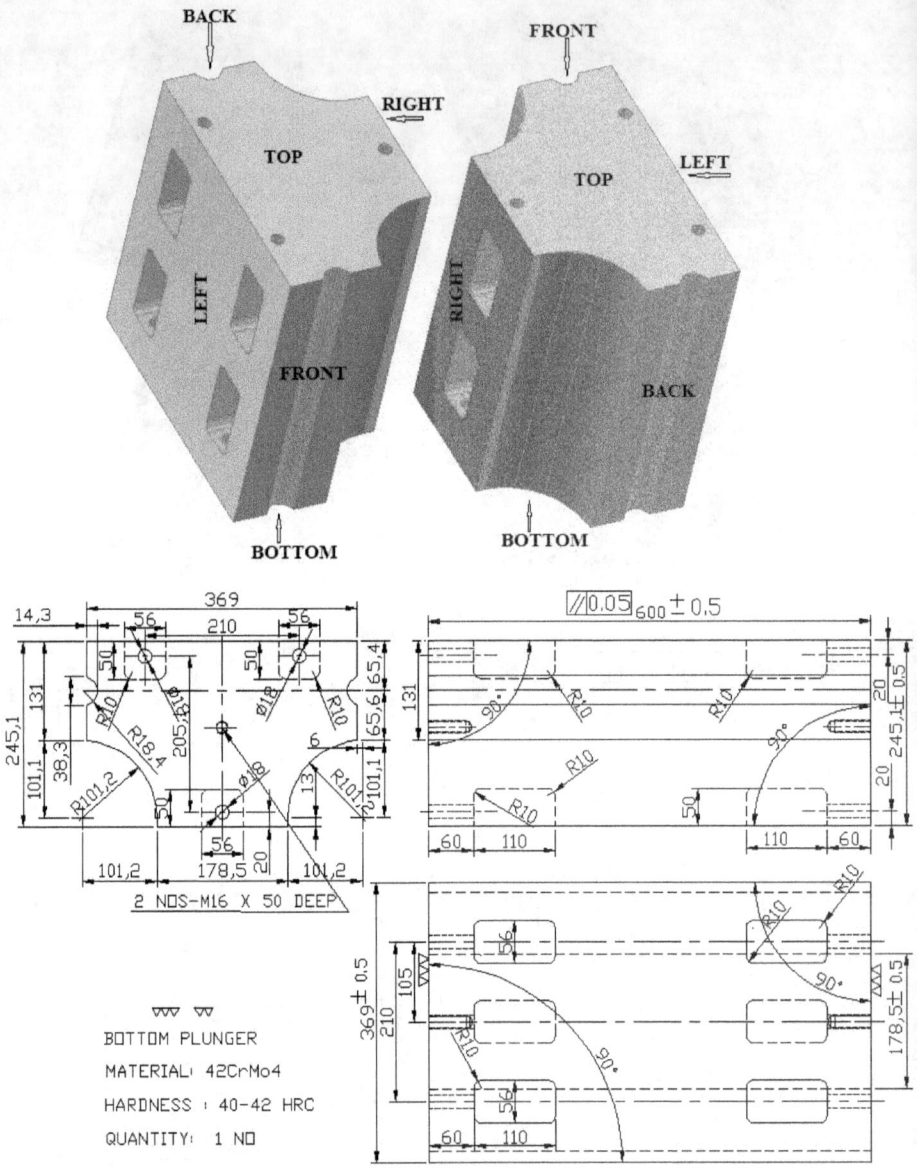

3D View of Bottom plunger:

The bottom die will be attached with Allen bolt to a face marked "TOP" and bottom plunger holding plate to the bottom face. The right side view has been rotated to show pockets and drill holes.

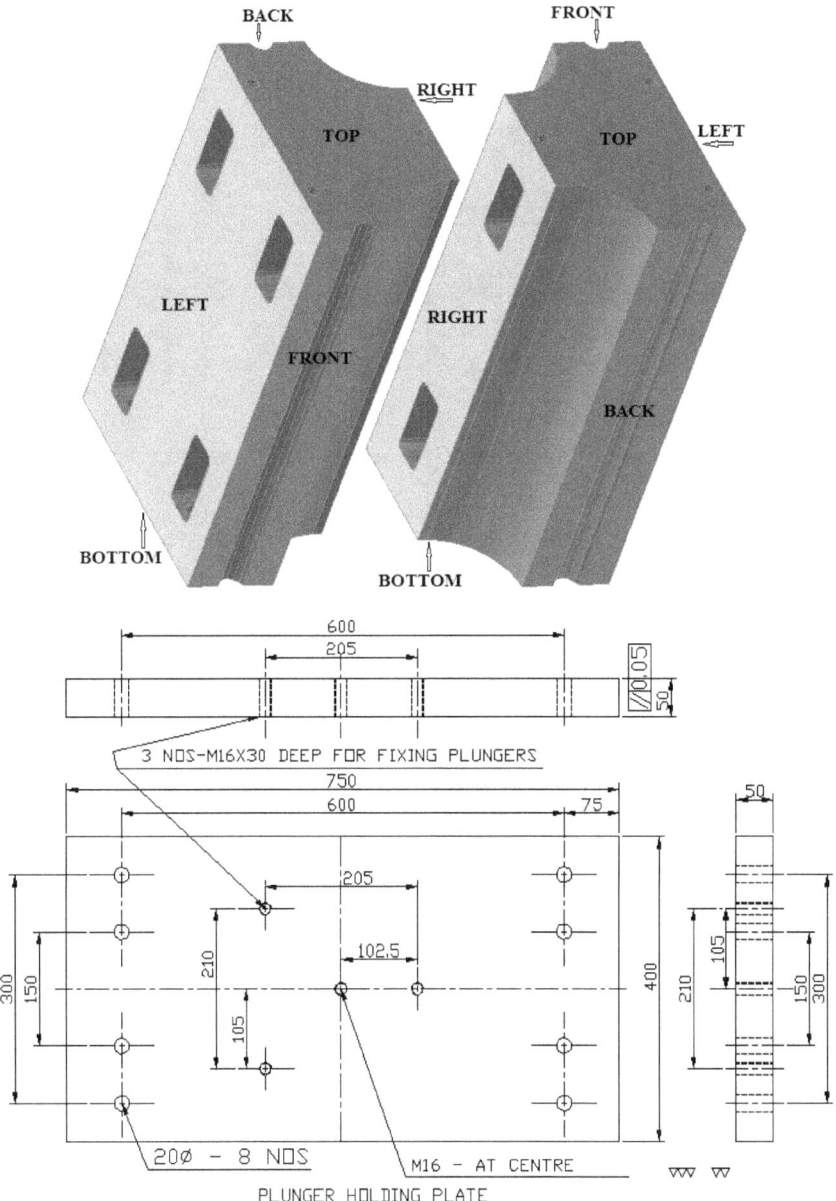

3 NOS-M16X30 DEEP FOR FIXING PLUNGERS

600
205
50
⌀0.05

750
600
75
205
102.5
300
150
210
105
400
210
105
50
105
150
300

20⌀ - 8 NOS

M16 - AT CENTRE

PLUNGER HOLDING PLATE

MATERIAL: 42CrMo4

HARDNESS : 40-42 HRC

QUANTITY: 2 NOS

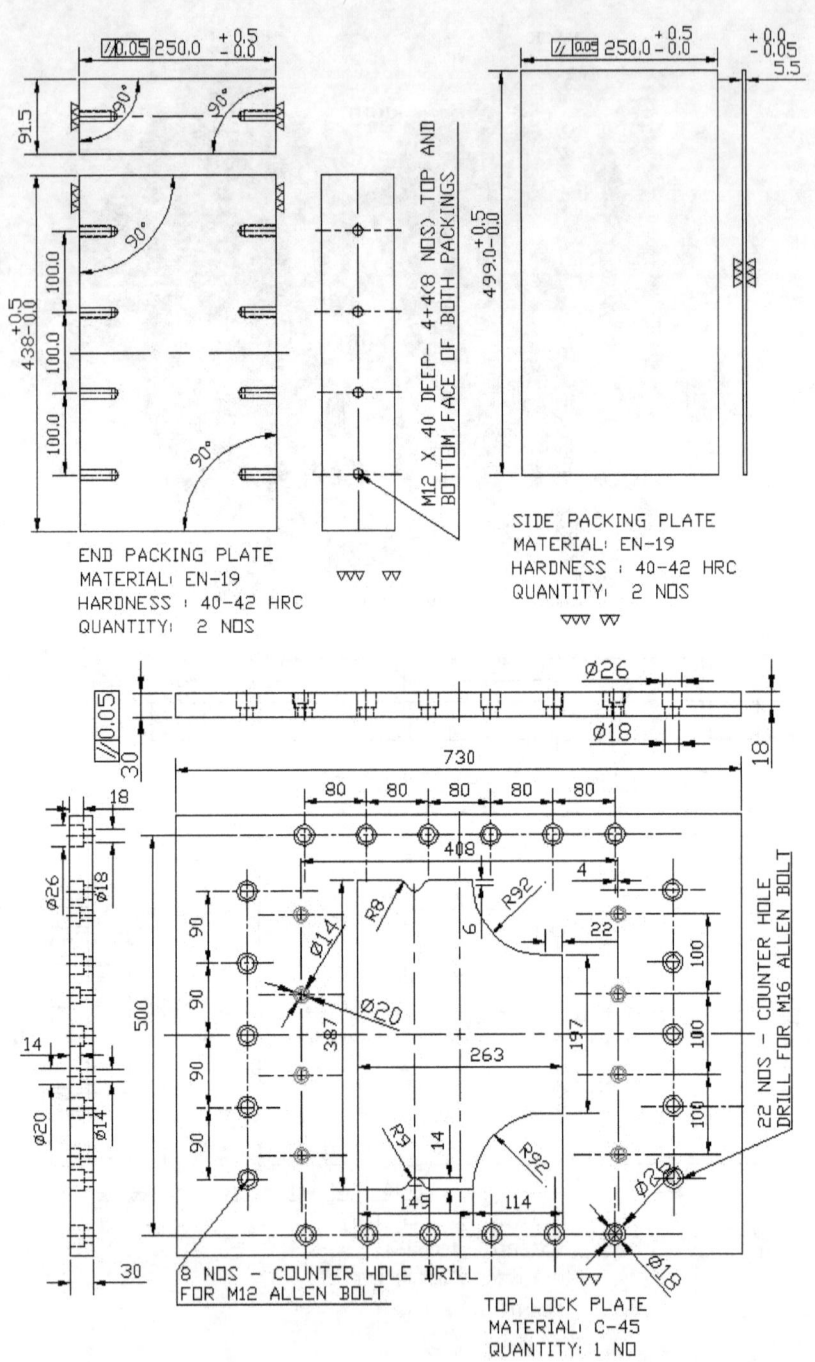

END PACKING PLATE
MATERIAL: EN-19
HARDNESS : 40-42 HRC
QUANTITY: 2 NOS

M12 X 40 DEEP- 4+4(8 NOS) TOP AND
BOTTOM FACE OF BOTH PACKINGS

SIDE PACKING PLATE
MATERIAL: EN-19
HARDNESS : 40-42 HRC
QUANTITY: 2 NOS

22 NOS - COUNTER HOLE
DRILL FOR M16 ALLEN BOLT

8 NOS - COUNTER HOLE DRILL
FOR M12 ALLEN BOLT

TOP LOCK PLATE
MATERIAL: C-45
QUANTITY: 1 NO

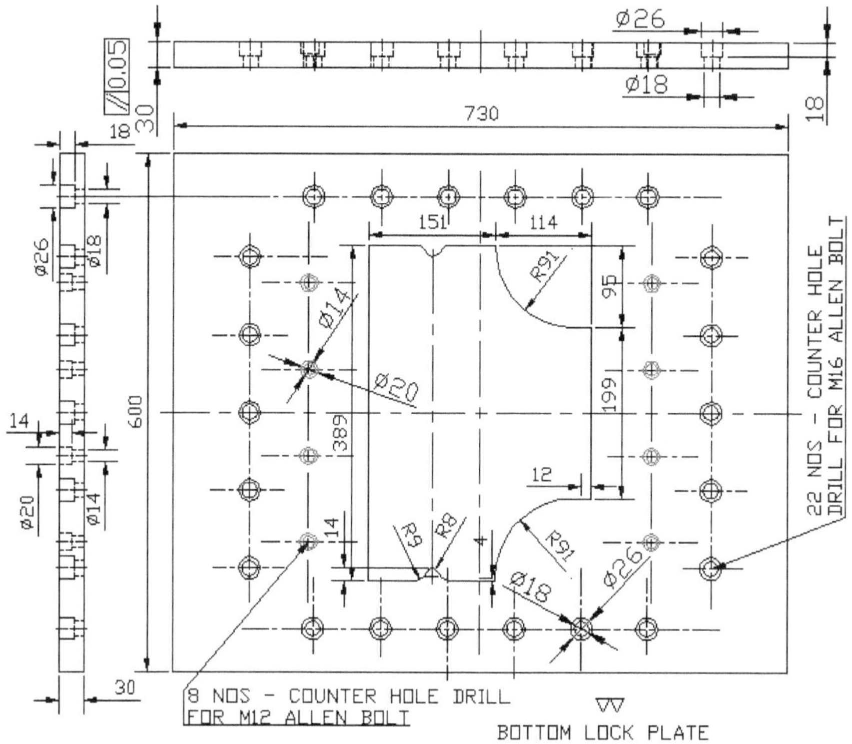

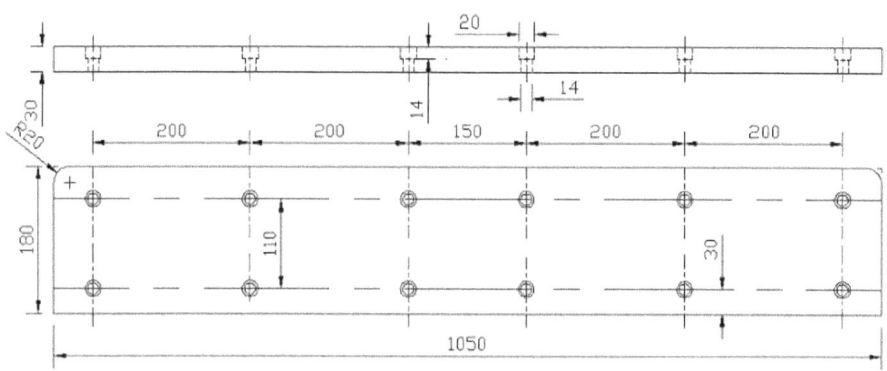

Ø26

Ø18

//0.05

30

18

730

18

151   114

R91

95

199

Ø14

Ø20

389

12

600

14

R9 R8

4

R91

Ø26

Ø18

22 NOS - COUNTER HOLE
DRILL FOR M16 ALLEN BOLT

Ø26

Ø18

14

Ø20   Ø14

30

8 NOS - COUNTER HOLE DRILL
FOR M12 ALLEN BOLT

∇∇

BOTTOM LOCK PLATE
MATERIAL: C-45
QUANTITY: 1 NO

20

14

14

R30
R20

200   200   150   200   200

180

110

30

1050

TOP COVER PLATE FRONT AND BACK

∇∇

MATERIAL: C-45
QUANTITY: 2 NOS

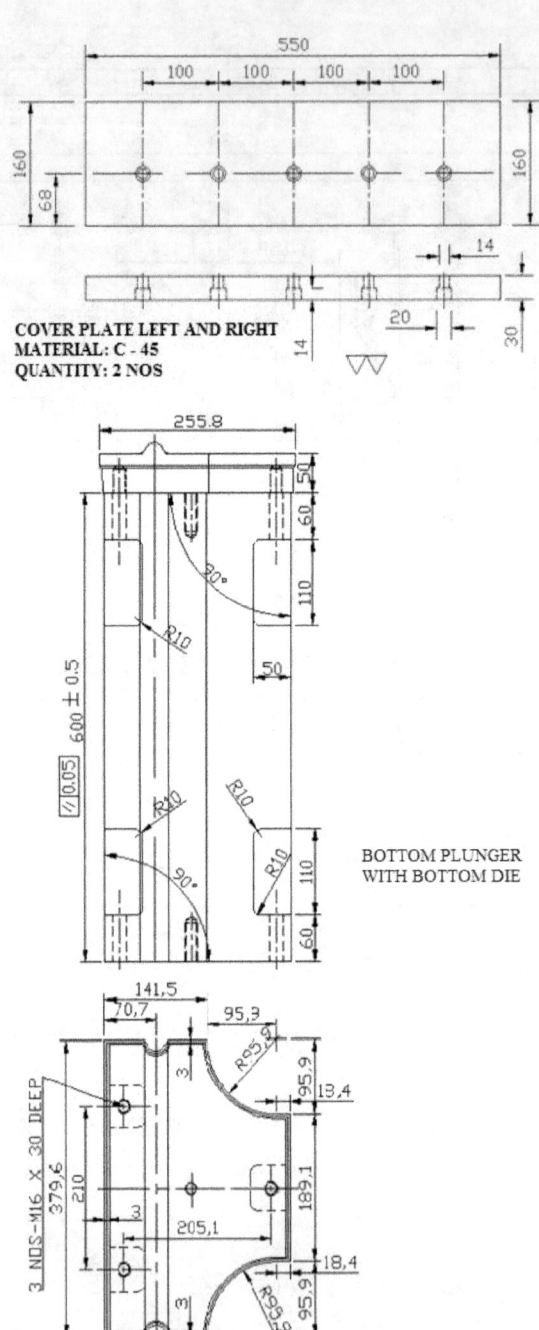

COVER PLATE LEFT AND RIGHT
MATERIAL: C - 45
QUANTITY: 2 NOS

BOTTOM PLUNGER
WITH BOTTOM DIE

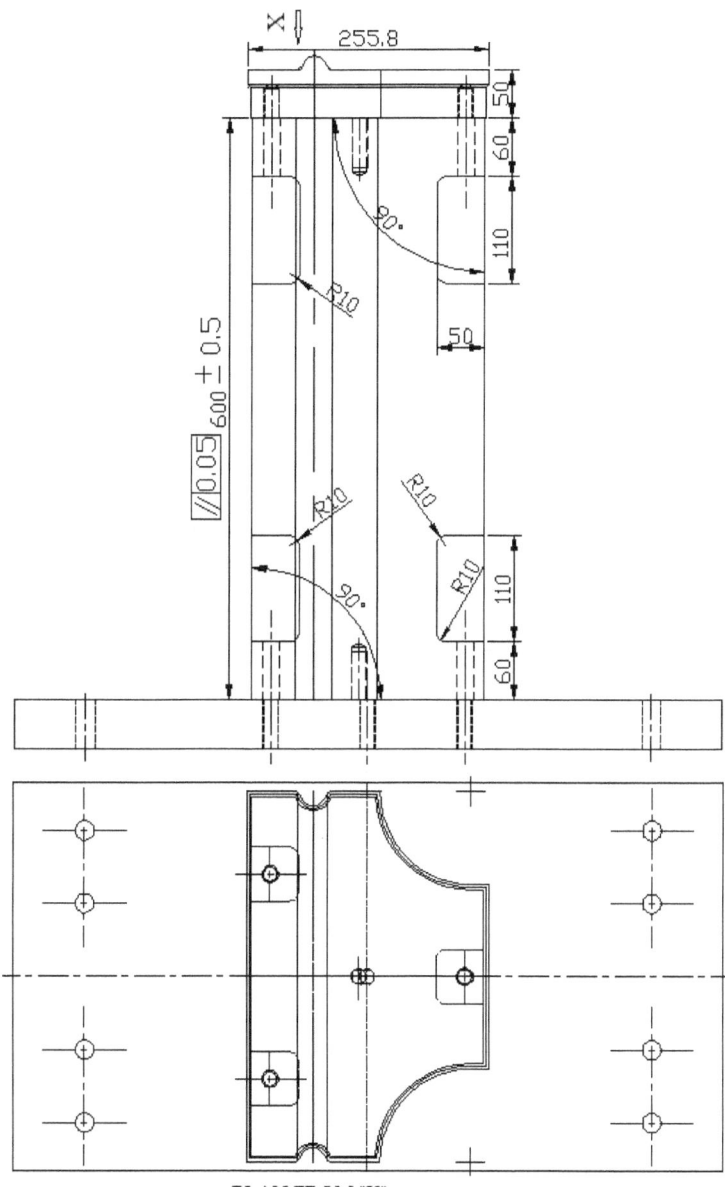

PLAN FROM "X"
BOTTOM PLUNGER, DIE AND PLUNGER HOLDING PLATE

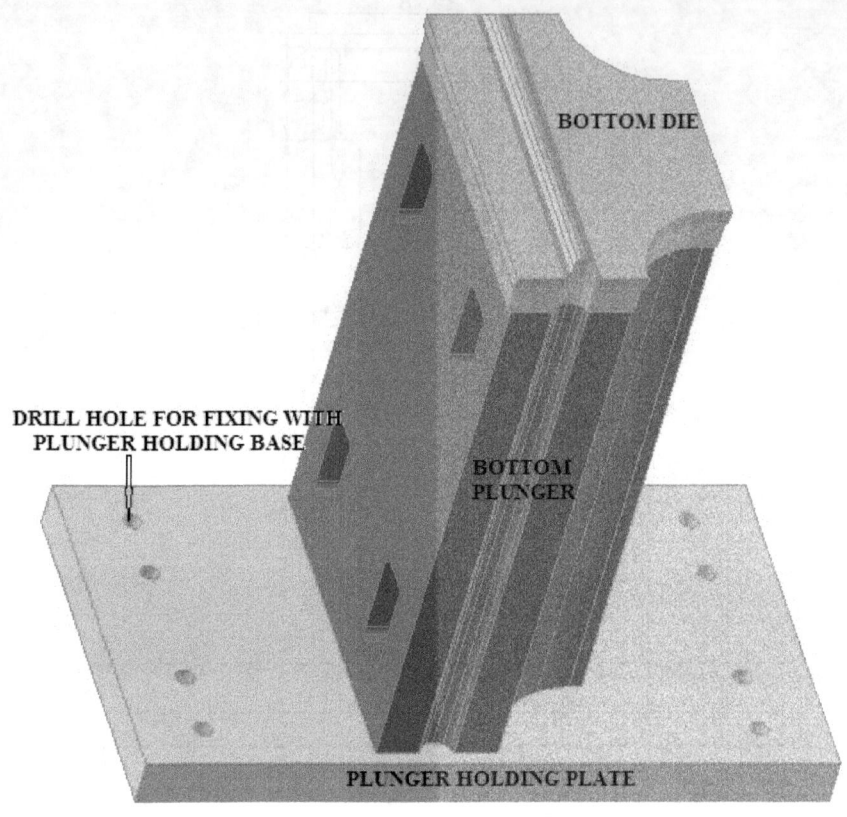

BOTTOM DIE

DRILL HOLE FOR FIXING WITH
PLUNGER HOLDING BASE

BOTTOM
PLUNGER

PLUNGER HOLDING PLATE

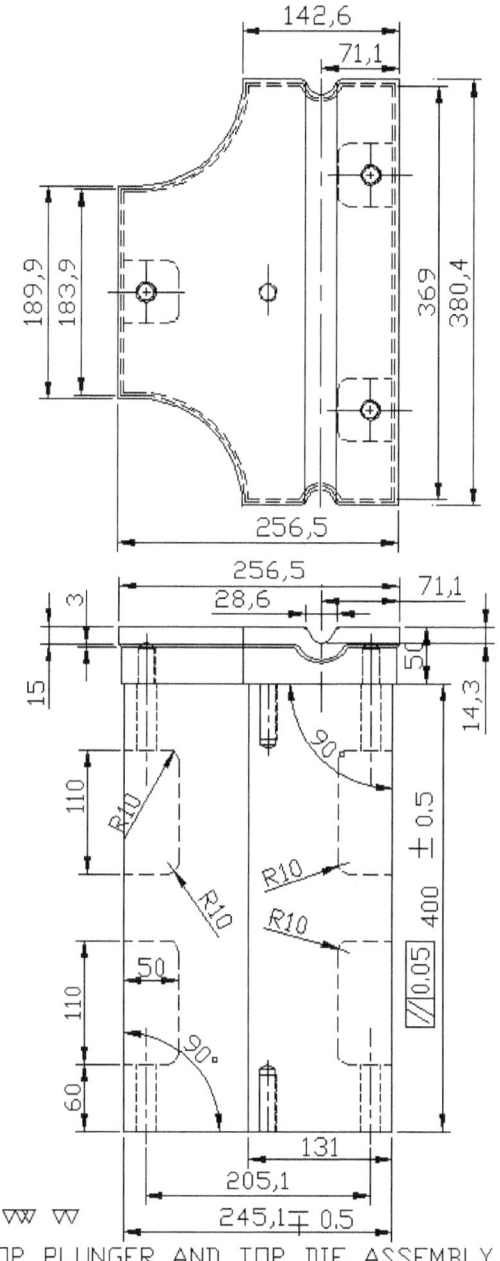

TOP PLUNGER AND TOP DIE ASSEMBLY

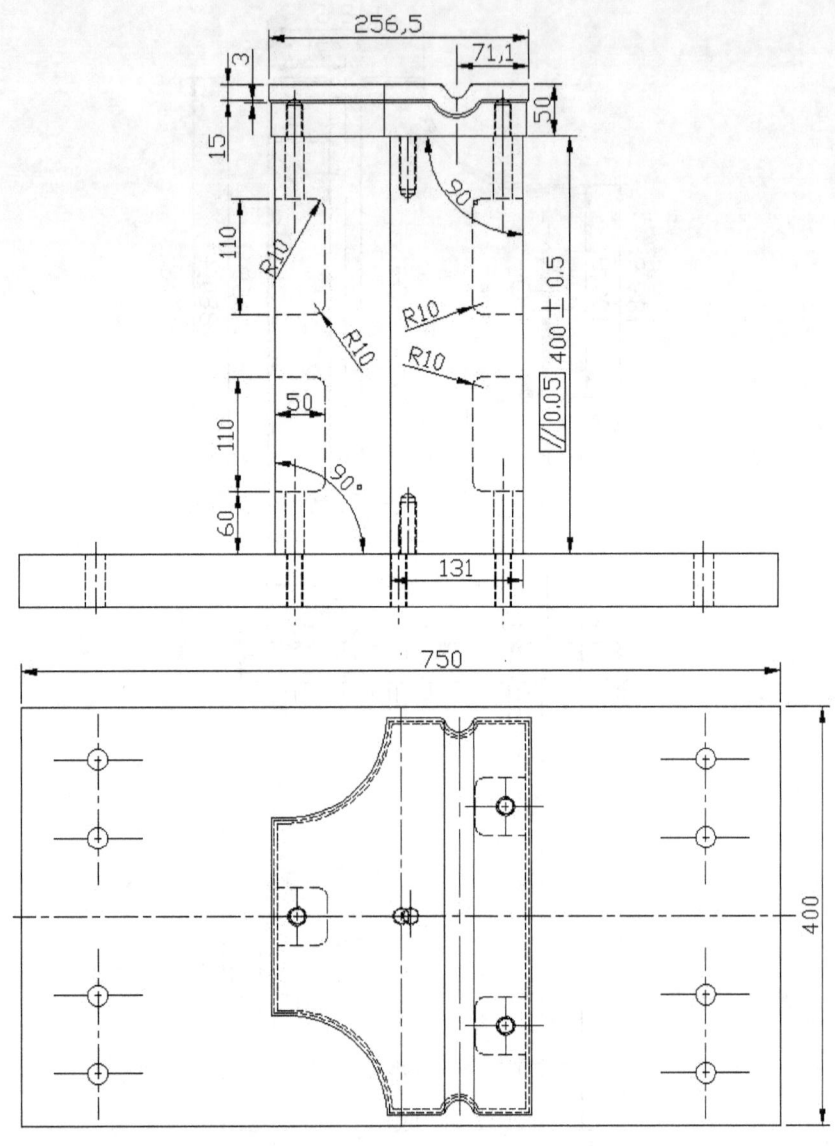

TOP DIE, TOP PLUGER AND HOLDING PLATE

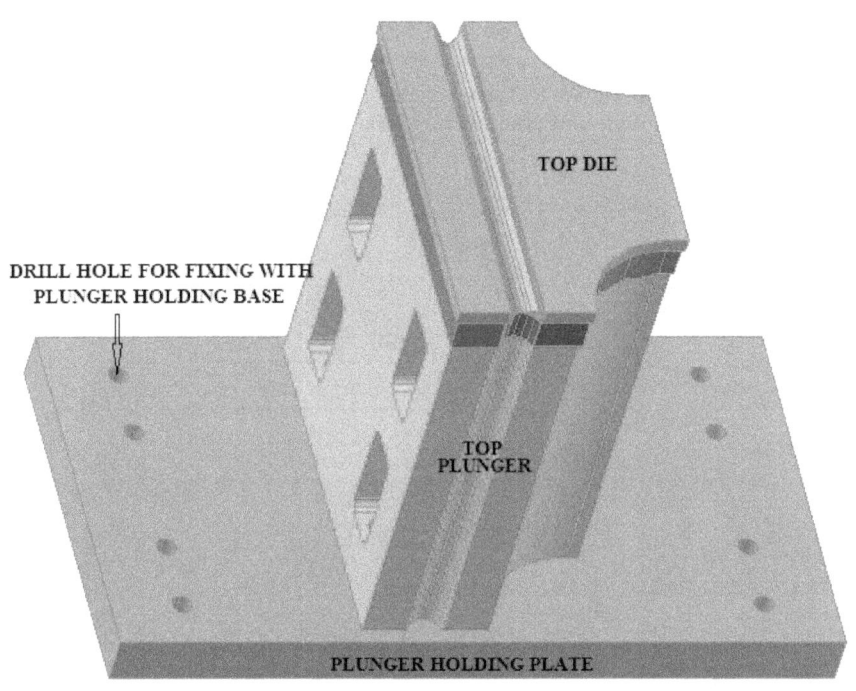

DRILL HOLE FOR FIXING WITH
PLUNGER HOLDING BASE

TOP DIE

TOP
PLUNGER

PLUNGER HOLDING PLATE

Assembly of liners with side and end packings: The front view has been drawn after removing EL-2, Lp-2 and side packing plate. The pressed brick is in between top and bottom dies. In the plan, the top die has been removed to show the brick shape and its location. The side views only brick location has been marked without dies.

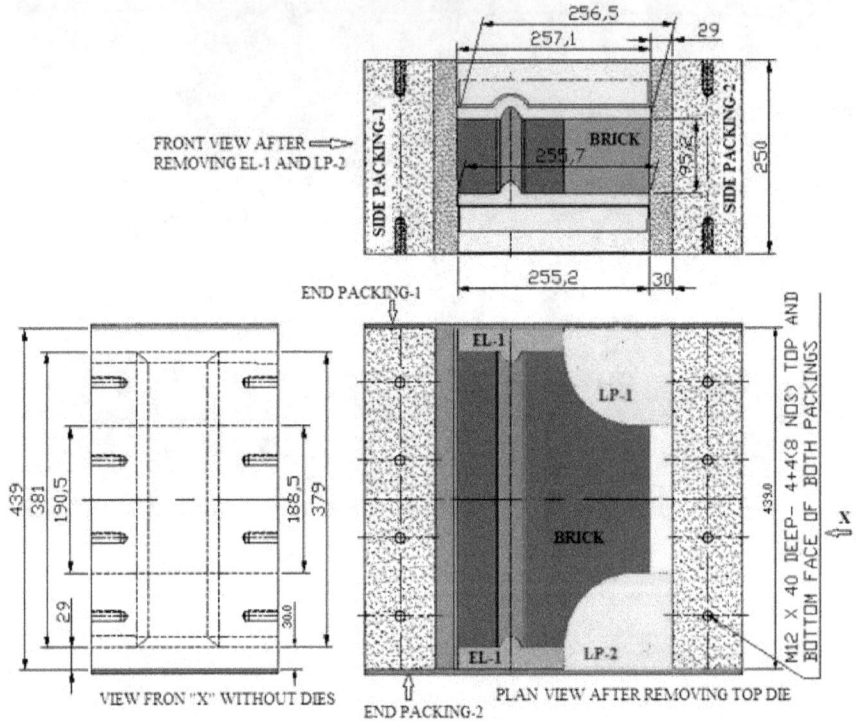

The liners and Packing assembly in mother mould:

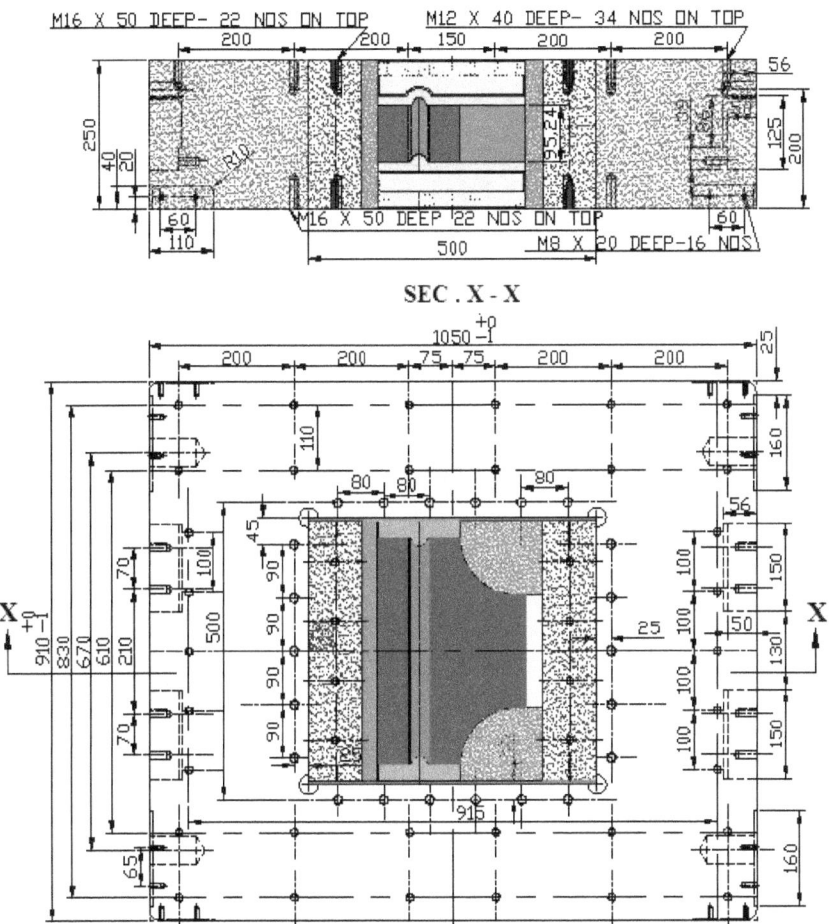

SEC . X - X

Liners, packing, mother mould and lock plate assembly:

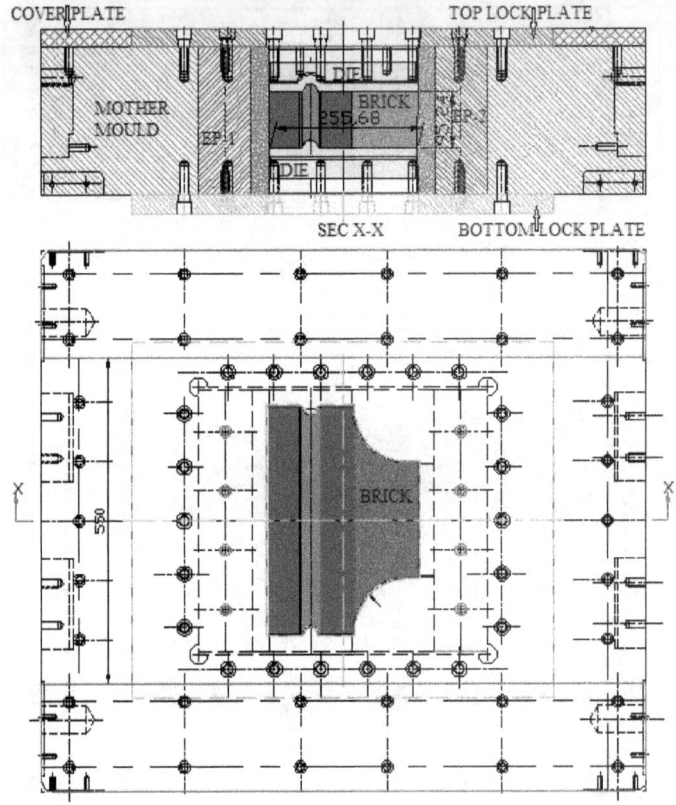

Liners and packing plate assembly in mother mould:

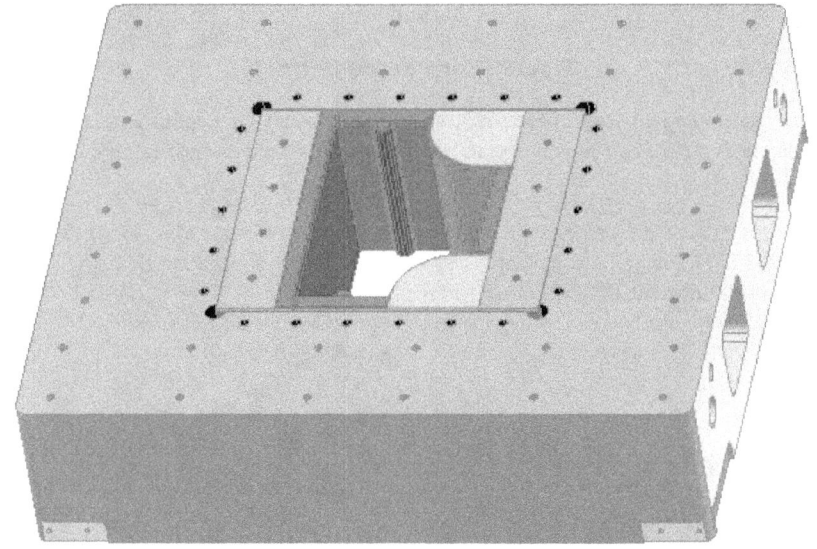

Liners, loose pieces, packing, bottom lock plate and top lock plate assembly:

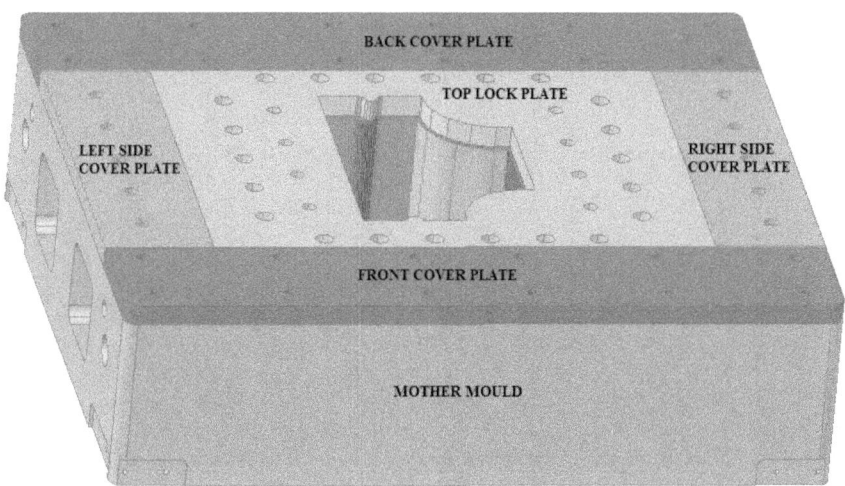

Chapter – 17

Mould design for dome brick

Note: The shape and sizes of the brick given here are not for application. It has been developed and designed just to explain the process of mould design and to educate the learners.

Let us design a mould for Dome brick that will neither expand nor shrink on firing. Design mould for 1000 Tonnes press having sliding table, fix bottom and top pressing system under the following condition for given shape and size.

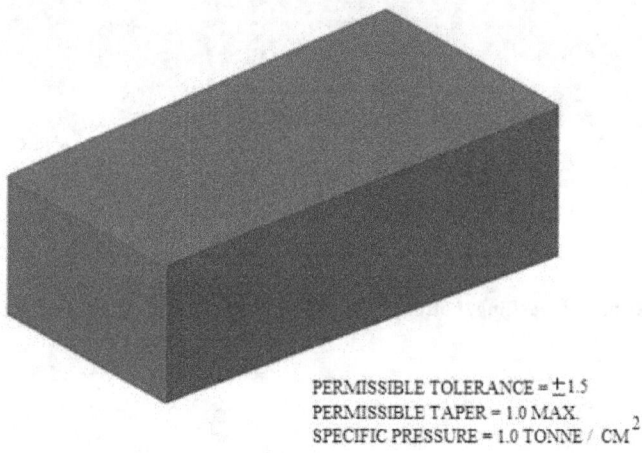

PERMISSIBLE TOLERANCE = ± 1.5
PERMISSIBLE TAPER = 1.0 MAX.
SPECIFIC PRESSURE = 1.0 TONNE / CM^2

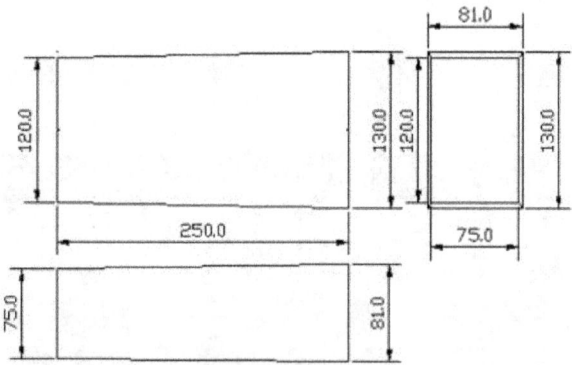

Make an overall plan to design the mould. First of all, calculate no of cavities that can be made with required specific pressure.

The open area of mould cavity face will be 120 / 130 X 250 as marked in front view.

The area of die face = 250*{(130 +120)/2} = 312.5 square cm

Absolute cavity size for 1.0 tonne per sq. cm specific pressure = 1000 / 1.0 = 1000 sq. cm

Number of cavities = 1000 / 312.5 = 3.2Cavities

We can make 3 cavities mould having 250 mm length along width of cavity in the mother mould.

Overall plan for mould design:

The two options have been planned for mould design as plotted in the drawings. The end liners and side liners in both options have been marked. The manufacturing of side liner and end liner in option -1 is easy, but the accuracy of assembly depends on the skill of man. It is difficult to have the same size in all three cavities. The liners may shift if there is uneven feeding of the mixture in the cavities. It is suitable for the single cavity.

In option-2 the side liner has undercut with the draft on the surface and the end face of the slot. The grinding of liners needs care and accuracy. The assembly will be more accurate and easy.

All the accessories have been designed for option -2

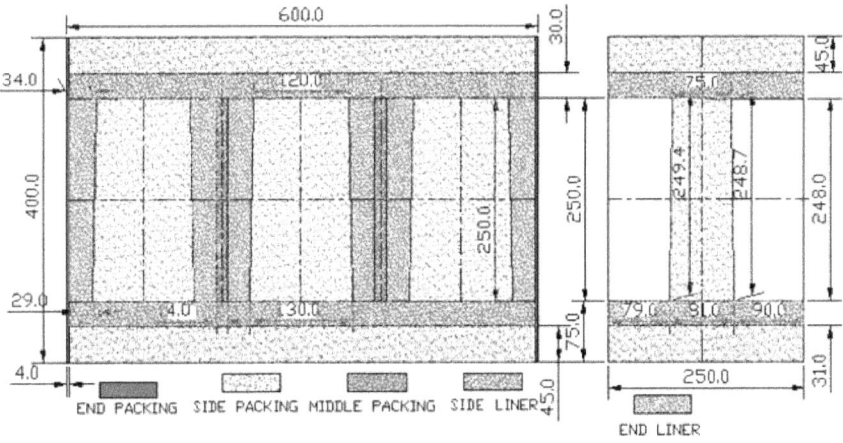

OPTION-1

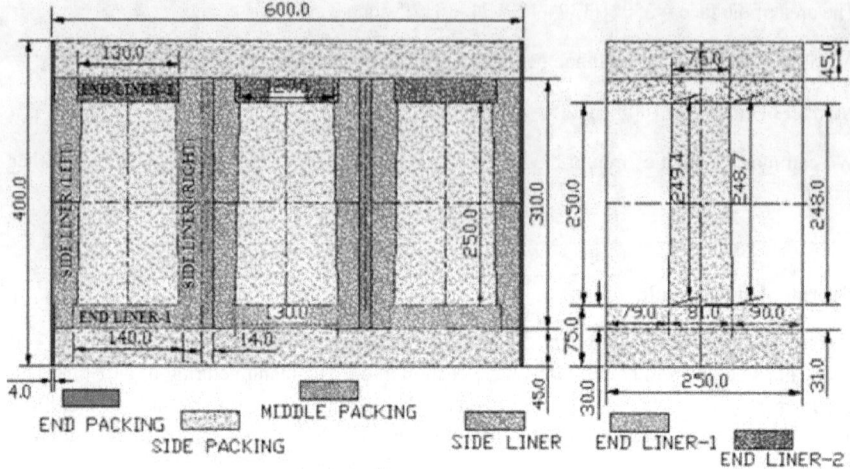

The mould accessories for this shape will be

- Mother mould
- Sideliners
- End liners
- Top die
- Bottom die
- Top plunger
- Bottom plunger
- Top plunger holding plate
- Bottom plunger holding plate
- Bottom lock plate
- Top lock plate
- Cover plate

Mother mould: To illustrate the process 1000 tonnes capacity press with top pressing and mould sliding table has been considered. The working parameter and process of mother mould design, the decision on numbers of the cavity have been explained earlier. The three cavities mould design on this press with 1.0 tonne/ per sq.cm specific pressure is possible as the area of die face will suit for three cavities. The mother mould with 600 X 400 cavity size and 250 heights will be suitable for moulding this brick. (See Chapter – 13 End Arch brick serial no 1). The design of tap holes for lock plates has been explained in mould design for End Arch brick.

The design of mother mould with cavity size and other details are given here.

The cavity size in mother mould can change according to the formation of the group, brick sizes and number of cavities. It has been explained in previous chapters.

The number of tap holes, location and centre to centre distance also may change according to length and width of the cavity. The radius of circular undercut at four corners of the cavity is 16 mm. The tap hole at corners for fitting lock plates should not be very close to it.

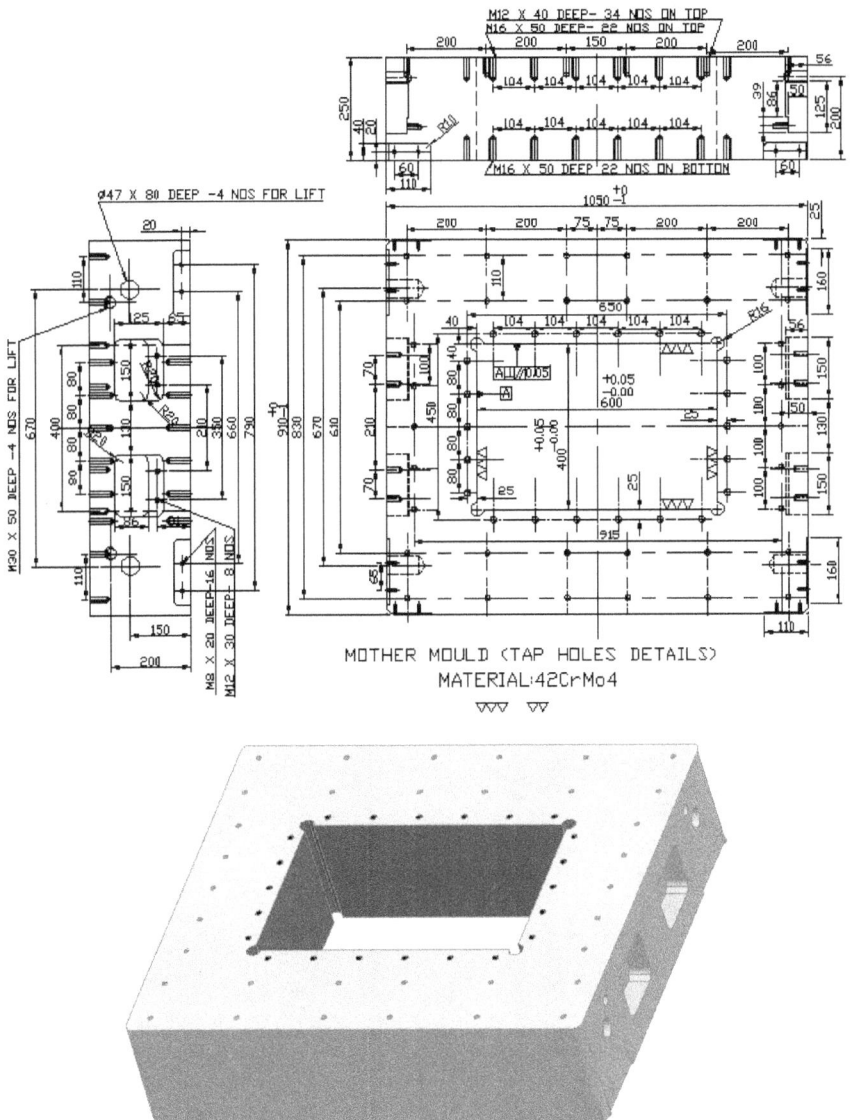

MOTHER MOULD (TAP HOLES DETAILS)
MATERIAL:42CrMo4

Side liners:

The left and right side liners are not symmetrical. It differs in the draft that is provided for ejection of bricks from the mould cavity. The thicknesses at both ends are not same, as such left side liner can't be fitted on the right side by reversing. The inclination on the inner face of mould cavity must be from bottom to top on both sides to have an equal draft on both sides. The person concern has to study both drawings given below very carefully to eliminate the chances of making wrong liners. The enlarged view is plotted here to visualise the two lines that represent draft.

Side liner (Left):

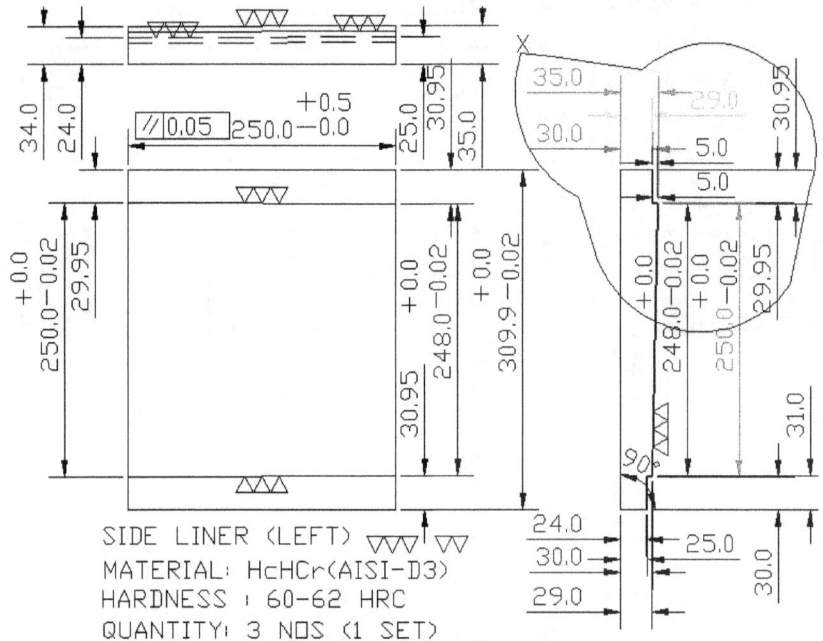

SIDE LINER (LEFT)
MATERIAL: HcHCr(AISI-D3)
HARDNESS : 60-62 HRC
QUANTITY: 3 NOS (1 SET)

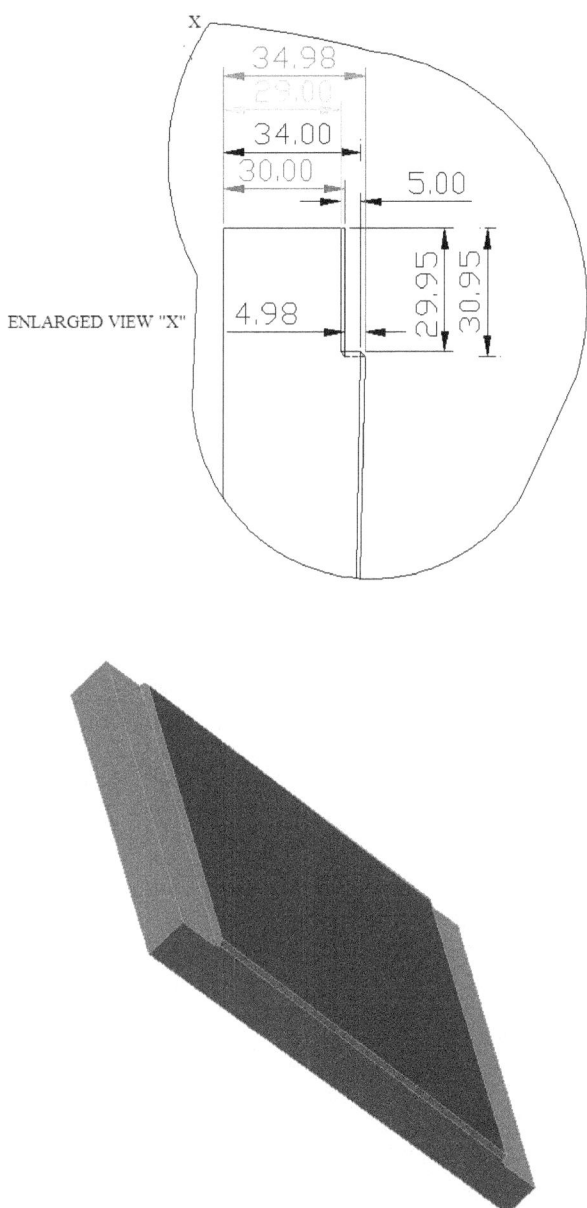

X

34.98

29.00

34.00

30.00

5.00

ENLARGED VIEW "X"

4.98

29.95

30.95

Side liner (Right):

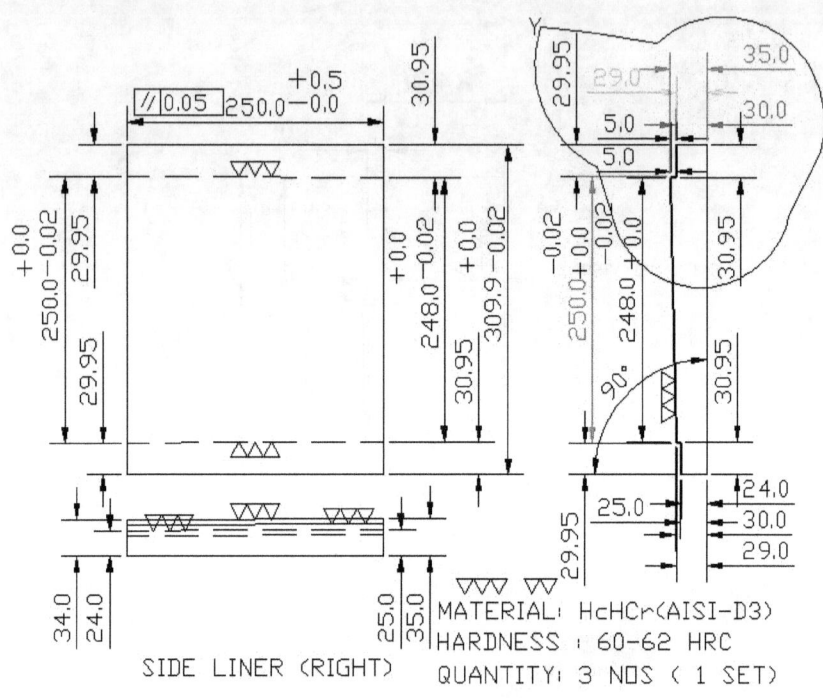

SIDE LINER (RIGHT)

MATERIAL: HcHCr(AISI-D3)
HARDNESS : 60-62 HRC
QUANTITY: 3 NOS (1 SET)

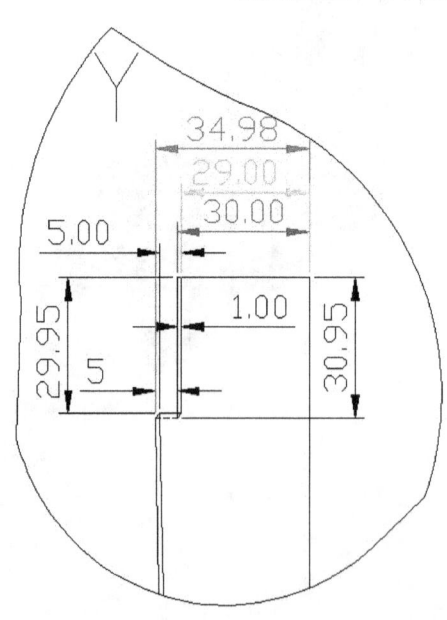

End liners:

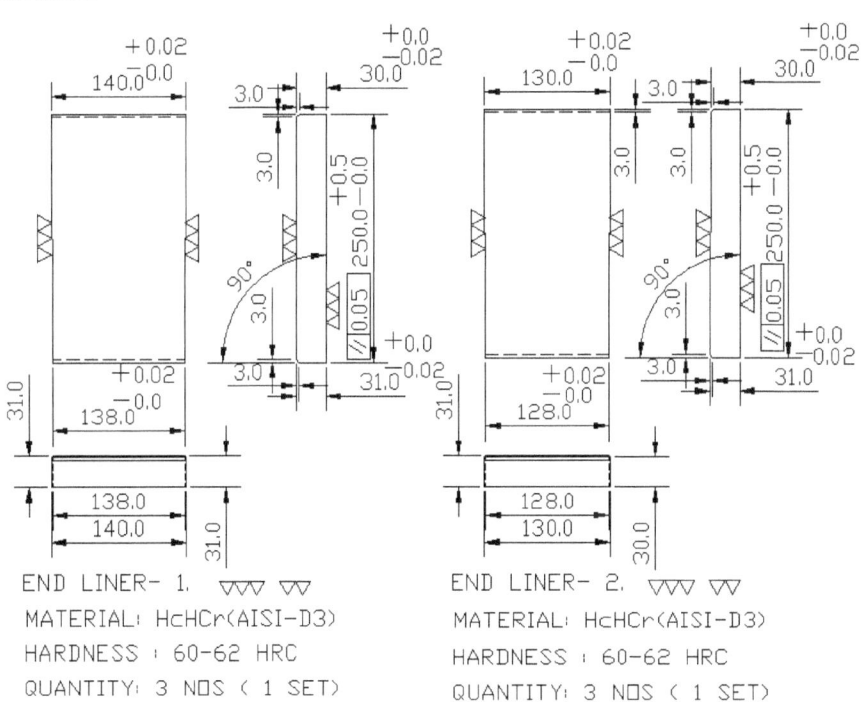

END LINER- 1. ▽▽▽ ▽▽
MATERIAL: HcHCr(AISI-D3)
HARDNESS : 60-62 HRC
QUANTITY: 3 NOS (1 SET)

END LINER- 2. ▽▽▽ ▽▽
MATERIAL: HcHCr(AISI-D3)
HARDNESS : 60-62 HRC
QUANTITY: 3 NOS (1 SET)

The boundary lines of mould cavity do not form a rectangular shape. There is a size difference in front and back side of end liners. The measurements have been taken from overall mould design plan layout. It has been marked as End liner – 1 and End liner – 2

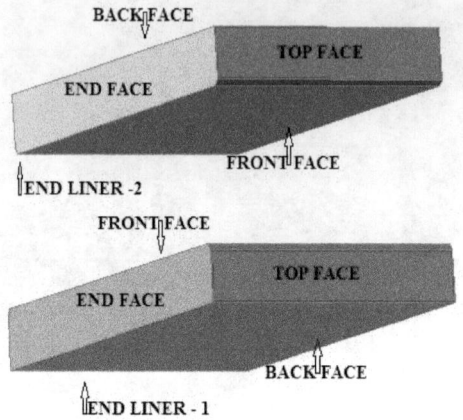

Side packing:

The thickness of side packing also has been taken out from overall mould design plan. The overall design plan has been made to suit the mother mould cavity size used for making End Arch brick. The tap holes on top and bottom face have been designed and plotted in the centre of thickness. The counter bore at matching location and size of Allen bolt to fit in these tap holes have been made in lock plates.

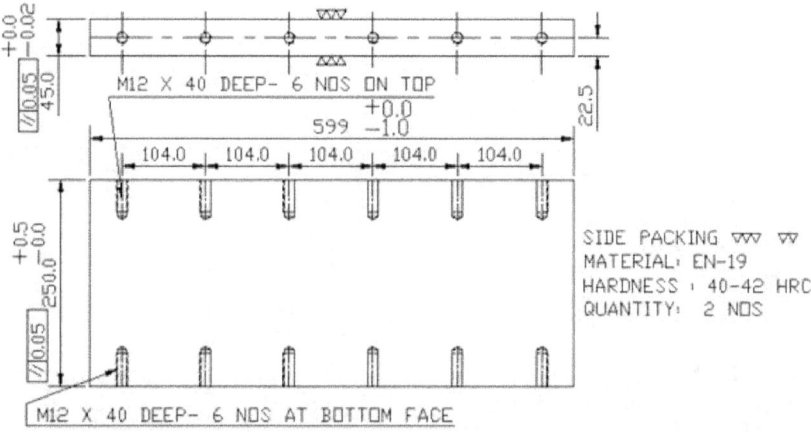

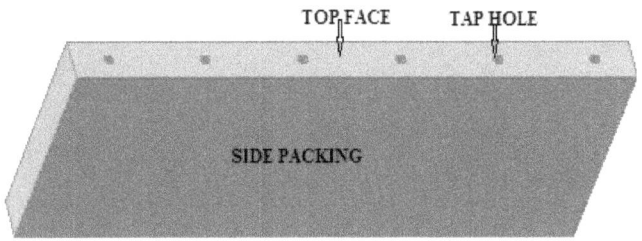

Middle packing:

The thickness of middle packing is decided considering gripper's arm thickness and its side way movement to hold the bricks. It should be 14 mm for the selected press configuration here to explain the design process. The gripper has to be modified to match with tapers in the sides of bricks.

End Packing:

The drawing of end packing also has been developed from overall mould design plan.

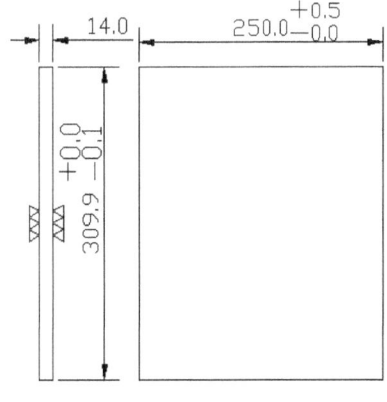

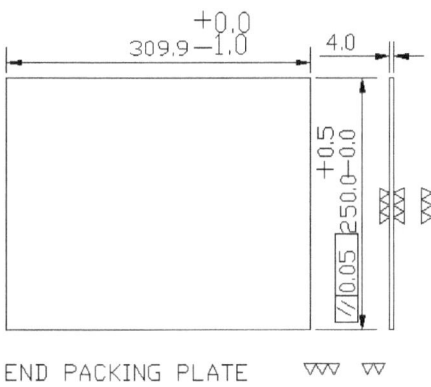

MIDDLE PACKING PLATE

MATERIAL: EN-19

HARDNESS : 40-42 HRC

QUANTITY: 2 NOS

END PACKING PLATE

MATERIAL: EN-19

HARDNESS : 40-42 HRC

QUANTITY: 2 NOS

Assembly of liners and packing:

The side liners and end liners with packing plates have been assembled to make 3 cavities mould.

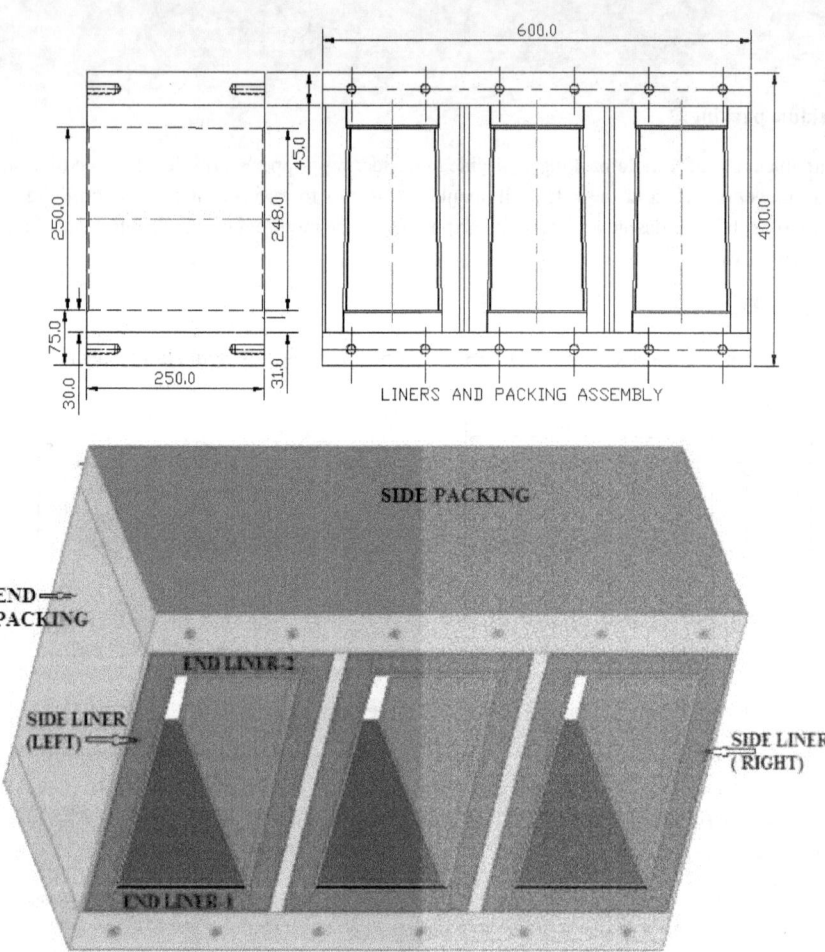

LINERS AND PACKING ASSEMBLY

Bottom Lock plate:

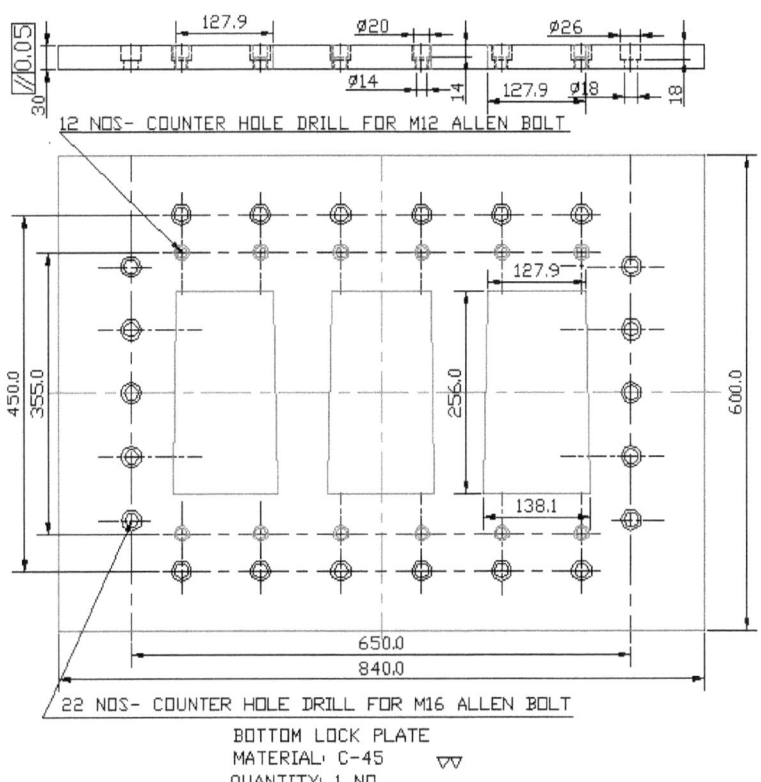

12 NOS- COUNTER HOLE DRILL FOR M12 ALLEN BOLT

22 NOS- COUNTER HOLE DRILL FOR M16 ALLEN BOLT

BOTTOM LOCK PLATE
MATERIAL: C-45
QUANTITY: 1 NO

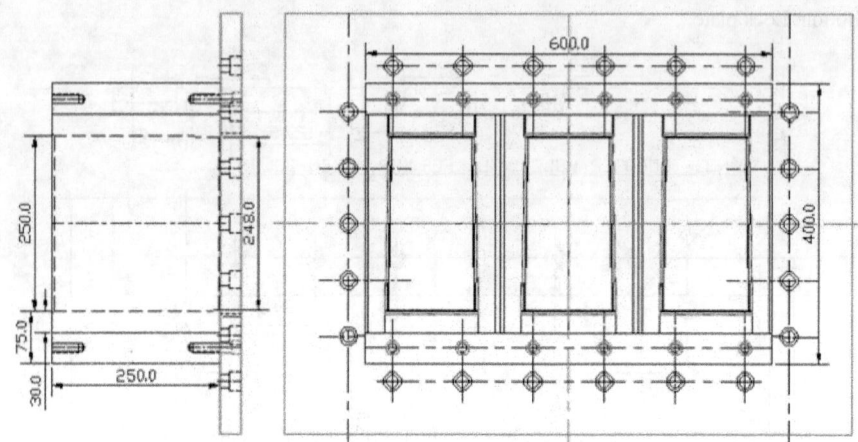

BOTTOM LOCK PLATE, LINERS AND PACKING PLATES ASSEMBLY

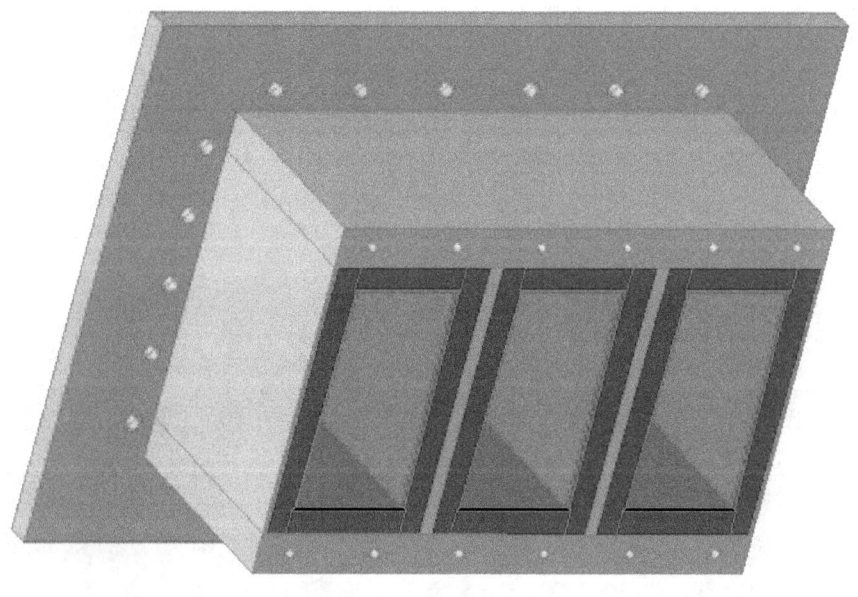

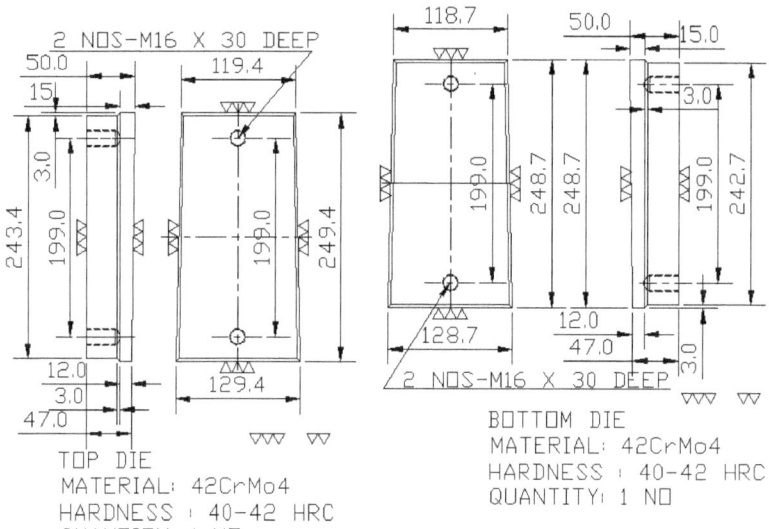

TOP DIE
MATERIAL: 42CrMo4
HARDNESS : 40-42 HRC
QUANTITY: 1 NO

BOTTOM DIE
MATERIAL: 42CrMo4
HARDNESS : 40-42 HRC
QUANTITY: 1 NO

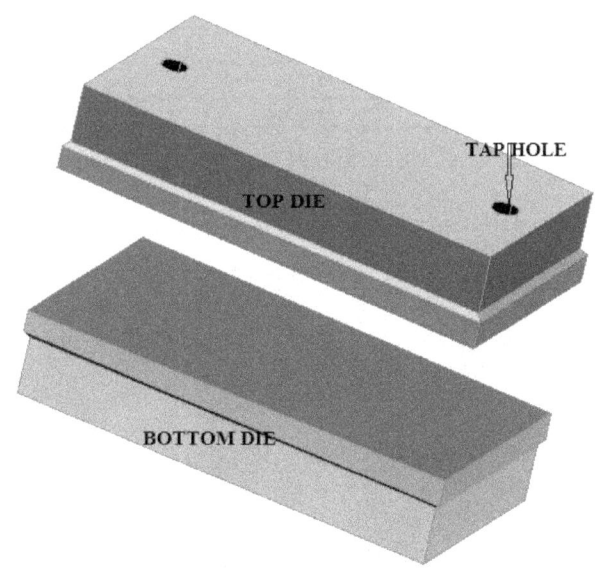

TAP HOLE

TOP DIE

BOTTOM DIE

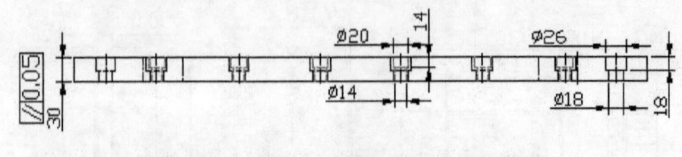

12 NOS- COUNTER HOLE DRILL FOR M12 ALLEN BOLT

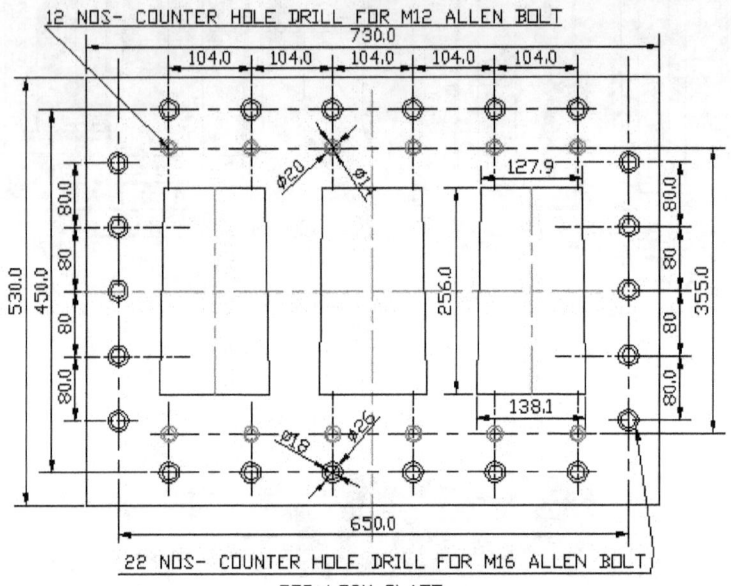

22 NOS- COUNTER HOLE DRILL FOR M16 ALLEN BOLT

TOP LOCK PLATE
MATERIAL: C-45
QUANTITY: 1 NO

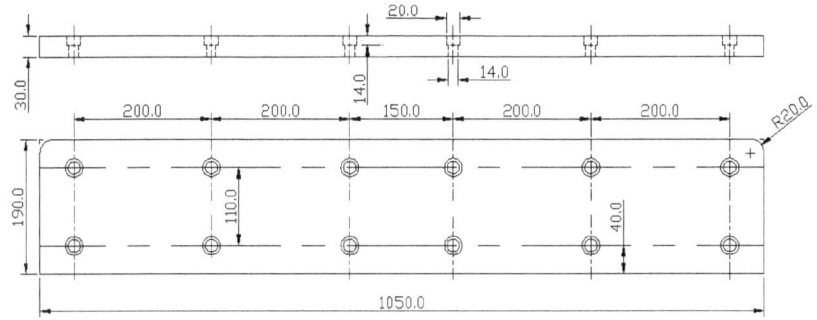

TOP COVER PLATE FRONT AND BACK
▽▽
MATERIAL: C-45
QUANTITY: 2 NOS

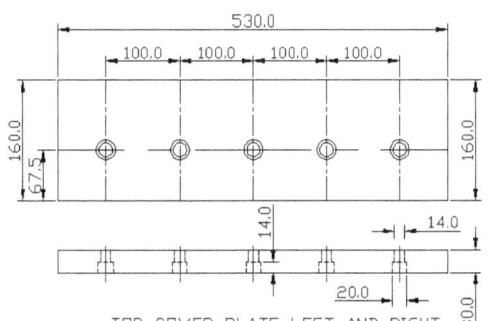

TOP COVER PLATE LEFT AND RIGHT
▽▽
MATERIAL: C-45
QUANTITY: 2 NOS

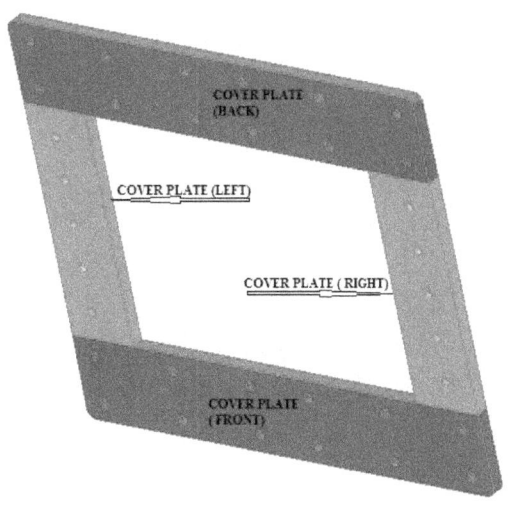

COVER PLATE (BACK)

COVER PLATE (LEFT)

COVER PLATE (RIGHT)

COVER PLATE (FRONT)

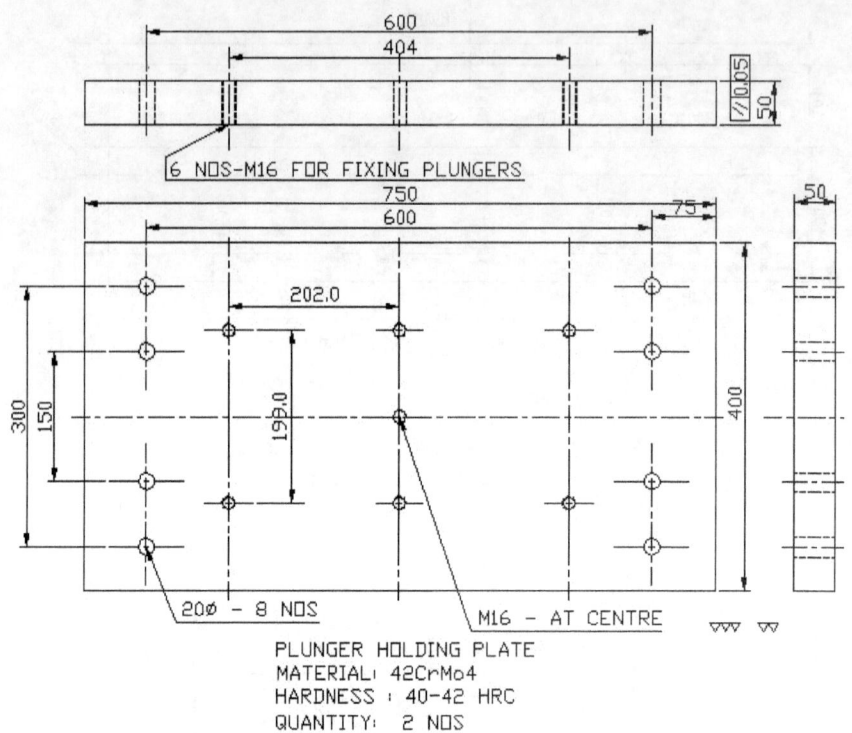

600
404

//0.05
50

6 NOS-M16 FOR FIXING PLUNGERS

750
600
75
50

202.0

300
150
199.0

400

20ø - 8 NOS

M16 - AT CENTRE

PLUNGER HOLDING PLATE
MATERIAL: 42CrMo4
HARDNESS : 40-42 HRC
QUANTITY: 2 NOS

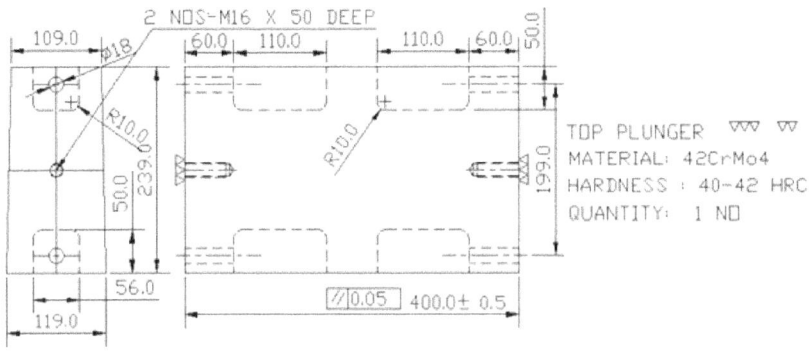

TOP PLUNGER ∇∇∇ ∇
MATERIAL: 42CrMo4
HARDNESS : 40-42 HRC
QUANTITY: 1 NO

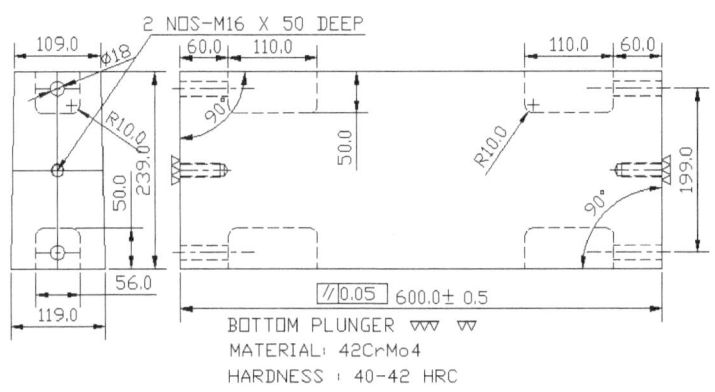

BOTTOM PLUNGER ∇∇∇ ∇
MATERIAL: 42CrMo4
HARDNESS : 40-42 HRC
QUANTITY: 1 NO

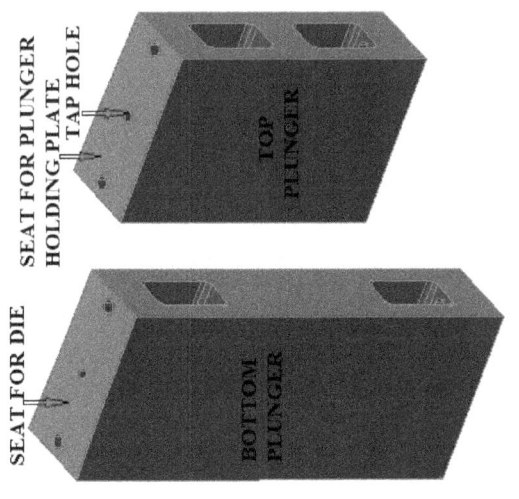

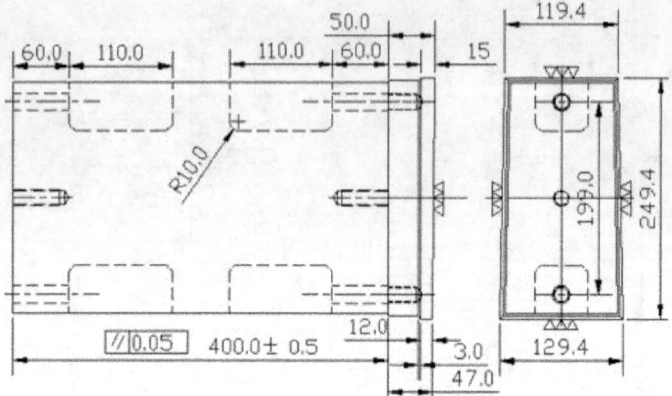

TOP PLUNGER AND TOP DIE ASSEMBLY

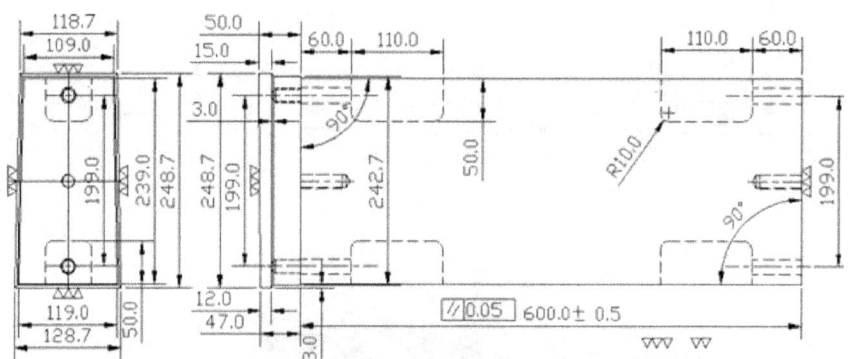

BOTTOM PLUNGER AND BOTTOM DIE ASSEMBLY

Top plunger, Top die and Plunger holding plate assembly:

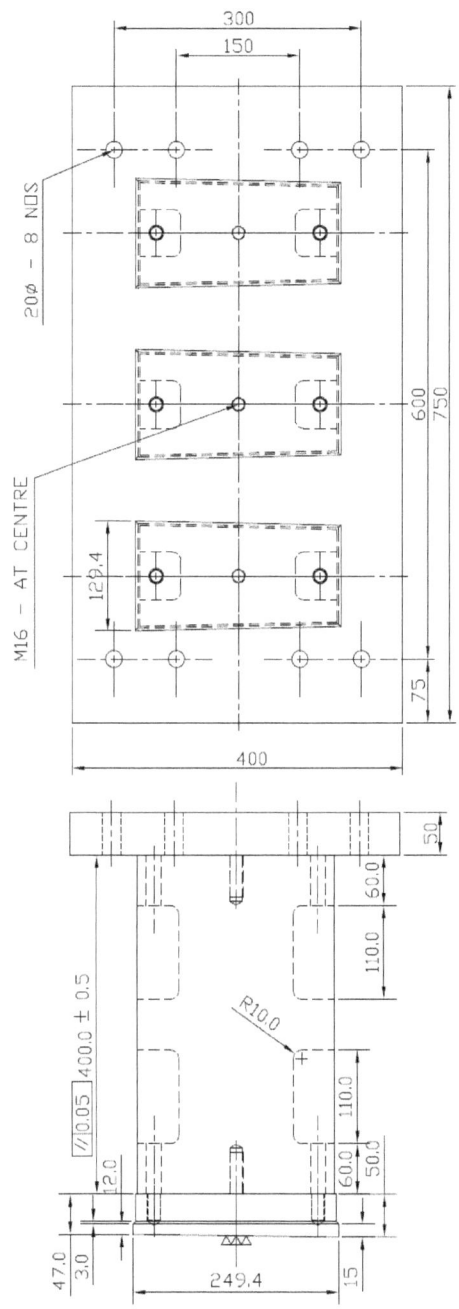

Bottom Plunger, Bottom die and Plunger holding plate assembly:

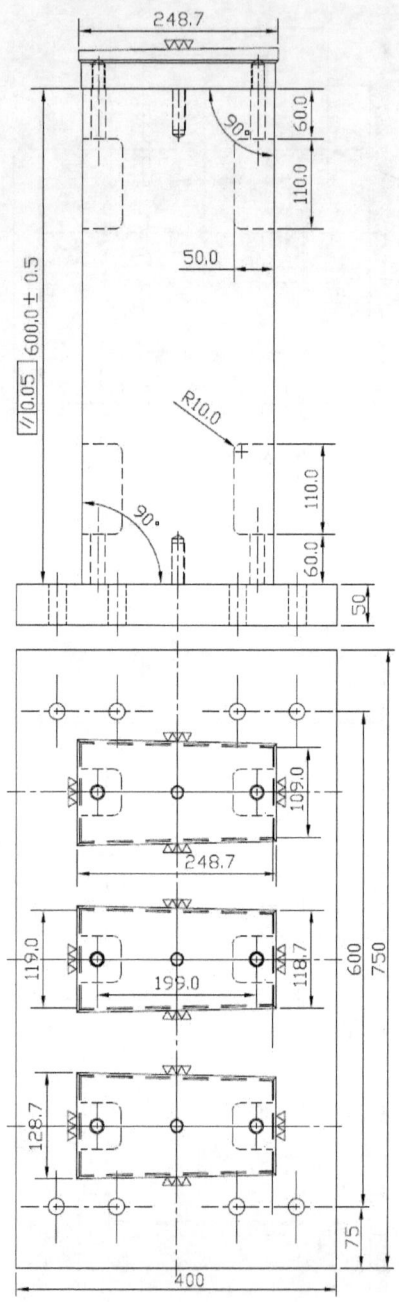

3D view of Plungers assembly with Dies and holding plates:

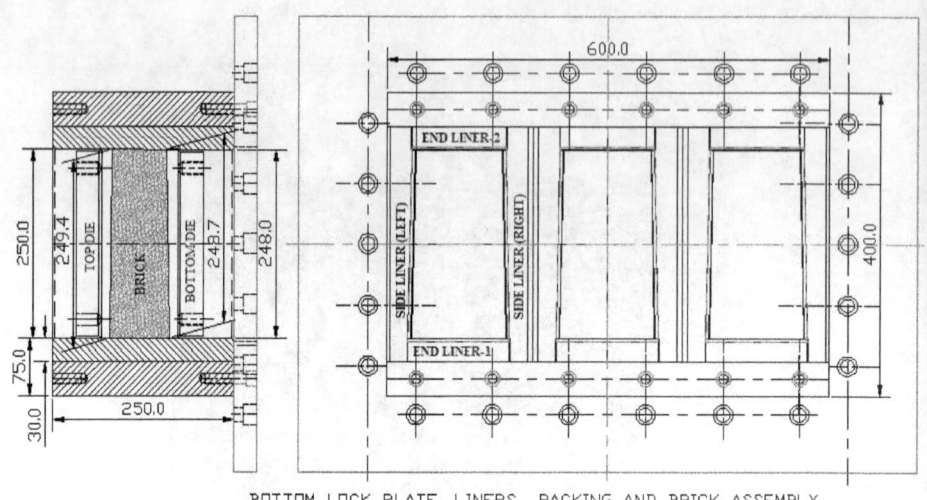

BOTTOM LOCK PLATE, LINERS PACKING AND BRICK ASSEMBLY

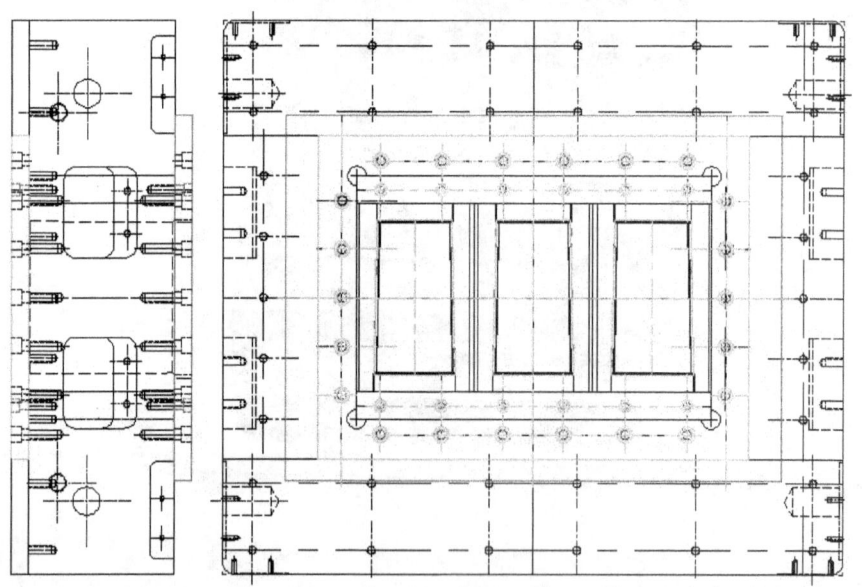

LINERS, PACKING, LOCKPLATES, COVER PLATES AND MOTHER MOULD ASSEMBLY

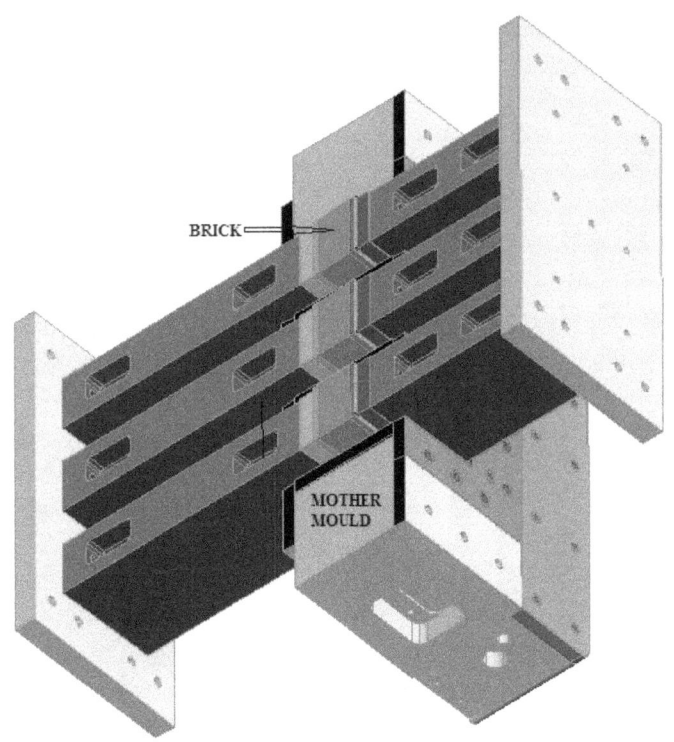

BRICK

MOTHER
MOULD

I would love to hear feedback and inputs on how I can make this more meaningful for readers and practitioners. You can provide your inputs and feedback by emailing at reesaa@indiavivid.com or visiting http://reesaa.com

-   Sheojee Prasad